Economics of Bioresources

Yoram Krozer · Michael Narodoslawsky
Editors

Economics of Bioresources

Concepts, Tools, Experiences

Springer

Editors
Yoram Krozer
Amsterdam, The Netherlands

Michael Narodoslawsky
Institute for Process and Particle Engineering
Graz University of Technology
Graz, Austria

ISBN 978-3-030-14617-7 ISBN 978-3-030-14618-4 (eBook)
https://doi.org/10.1007/978-3-030-14618-4

Library of Congress Control Number: 2019932830

This Springer imprint is published by the registered company Springer Nature Switzerland AG
The registered company address is: Gewerbestrasse 11, 6330 Cham, Switzerland

Preface

The *Economics of Bioresources* introduces economic perspectives on resources of biological origin. The essential property of bioresources is that they can be sustained infinitely as their disposal after use closes all material cycles, and new bioresources may be generated on fertile land using solar radiation. Bioresources are also essential for many daily uses, in particular nutrition, and can be replenished when nature is used with care. Unfortunately, those qualities are often neglected in economic thinking, though they contribute to welfare and business.

Valuable bioresources are addressed in this book. The purpose of this book is for students, scholars, managers, and policy makers to better understand the challenges and opportunities faced in the production, processing, and use of bioresources. Rather than elaborating on agricultural and forest-based production that deliver raw materials, the book is focused on industries and services as these add value to bioresources. The main theories in economics are introduced along with basic technologies and logistics, followed by practices in a few European countries. While we address biofuels and energy products based on bioresources, other products and services are introduced as well.

The idea for this book emerged during the BioEnergyTrain project. This was a project that developed academic education on bioresource engineering and management as these subjects were absent in the master's curricula. The master of biorefinery engineering has been implemented at the Graz University of Technology in Austria, and bioresources management is a specialization in the master of sustainable energy technologies at the University of Twente in the Netherlands. The editors of this book participated in that project, and other authors are practitioners in bioresources.

This volume has been realized with the support of many people. In particular, we are grateful to the European Sustainable Energy Innovation Alliance that coordinated the BioEnergyTrain project; of special importance were Brigitte Hasewend who led the project, and Wolfgang Bauer and Maarten Arensten who coordinated the development of the engineering and management masters, respectively.

We wholeheartedly appreciate several discussions that ensued on sustainable purchases of bioresources with Jan-Kees Vis of Unilever; local bioenergy uses with Neven Duic of the Technical University of Zagreb; and bioresource in regional services with Igor Studennikov and Oleg Dyakov of the Center of Regional Studies.

We are grateful to Shonali Chenzira who edited all texts in this book. Finally, our appreciations to two unknown referees who orchestrated the proposal for this book as well as to Fritz Schmuhl, Naomi Portnoy, Catalina Sava, and Ram Prasad Chandrasekhar of Springer who took the risk of publishing it.

The book is dedicated to Rob Donia, Former Director of Engineering of Unilever, who introduced the term 'downscaling of technologies' with reference to bioresources several decades before it appeared in policy making.

Amsterdam, The Netherlands Yoram Krozer
Graz, Austria Michael Narodoslawsky

Contents

Part III Reviews of the Sectors

Editors and Contributors

About the Editors

Yoram Krozer received an M.Sc. in biology and M.A. in economics at the Utrecht University, Business Administration at Inholland, and Ph.D. in economics at the University of Groningen. He worked at non-governmental organizations and businesses, thereafter. After 30 years in the private sector, he joined the University of Twente as Director of the Cartesius Institute, Institute for Sustainable Innovations of the Netherlands' Technical University. He works on economics of sustainable development as Associate Professor at the University of Twente, Visiting Professor at the Jyothi Institute of Technology in India, and Director of the Sustainable Innovations Academy.

Michael Narodoslawsky did his studies and Ph.D. in chemical engineering at the Graz University of Technology. Until his retirement in 2016, he headed the research group Process Synthesis, Process Evaluation and Regional Development at the Graz University of Technology. He headed the 'Bioenergy' Working Group within the Education Task Force of the Strategy Energy Technology of European Commission and currently co-chairs the working group of 'Bioresources' for the European Sustainable Energy Innovation Alliance (ESEIA). He has research experience in ecological assessment and life cycle analysis, regional technology networks for renewable resources, and biorefinery development.

Contributors

Sandor Bartha received an M.Sc. in chemical engineering at the Universitatea Politehnica Timişoara and Ph.D. in technical agricultural engineering science (2009) at the 'SZIU University—Godollo.' He has more than 30 years of experience in the development and design of systems based on new energy sources. He is Accredited Researcher in the fields of solar energy conversions systems and thermochemical conversion of biomass. Recently, as part of the Marie Curie research program—PHOENIX, he has further specialized in the field of biorefinery processes, developing new technology schemes for the processing of solid biomass for second- and third-generation biofuel production.

Shonali Chenzira is a specialist in creating and managing content. She is skilled in all facets of creative strategy, concept design and development, editing, creative direction, iterative field testing, writing and implementing problem-solving strategies on the ground. Her expertise has been honed in diverse fields—from conservation and ecology education to organizing independent music festivals, curriculum development and experiential education, advertising copywriting and technical editing, among others. She has led the development of national-level programs on snakebite mitigation and financial literacy for rural India. A freelancer based out of Bengaluru (Southern India), she can be reached at shonali.chenzira@gmail.com.

Karin Fazeni-Fraisl received a master's degree in social economics at the Johannes Kepler University Linz in Austria and holds a doctorate degree in social sciences, economics, and business. She started working as Scientific Co-Worker at the Energy Institute at Johannes Kepler University—a nonprofit research association—in 2007 and has continued researching at this institution until today. Currently, she is Senior Researcher focusing on life cycle analysis of emerging technologies.

Marlene Kienberger (maiden name Fritz) received a DI in chemical engineering and Ph.D. in process engineering at Graz University of Technology. After the Ph.D., she worked as Process Engineer and later Assistant Production Manager in the industry. Having completed five years in the private sector, she joined the Graz University of Technology as Assistant Professor and worked toward building up the master's program 'biorefinery engineering.' She is also involved with the working group at the Institute of Chemical Engineering and Environmental Technology in dealing with the topics of bioeconomy/biorefinery in terms of chemical engineering.

Willem van Laarhoven received his B.Sc. from the HAS University of Applied Sciences in Den Bosch, M.Sc. in ecology from the Utrecht University. For seven years, he worked for the Centre for Agriculture and Environment, a nonprofit organization in the field of sustainable livestock farming. In the decade between 1990 and 2000, he worked in the livestock industry. He then started his own consultancy firm—Valacon B.V.—advising private companies and Dutch provincial organizations on sustainable livestock production. Since 2006, he has focused on the economic and environmental impact of dairy cattle and their longevity.

Johannes Lindorfer earned his master's degree at the University of Applied Sciences in Wels, Austria, studying bio- and environmental technology. Since 2007, he has conducted research, and as of 2009, he has been Project Leader of the Energy Institute at Johannes Kepler University. Additionally, he is a doctorate student in chemical engineering at the Graz University of Technology. His research focuses on the techno-economic assessments of emerging technologies.

Maia Lordkipanidze (1971) obtained her M.A. in environmental management and policy at Lund University, IIIEE, in Sweden. She obtained Ph.D. at the University of Twente, CSTM, on 'governance for resilience in protected areas.' Prior to her Ph.D. research, she worked for the Cartesius Institute for Sustainable Innovations of the Netherlands' Technical Universities on tourism and innovations. She also worked for the Province of Friesland, undertaking research and developing project proposals on regional development and innovations. Currently, she works on researching ecosystem services at the University of Twente.

Robert Tichler earned his doctorate degree in economics. He is Project Manager at the Energy Institute and Lecturer in the course 'energy management.' Since 2013, he has been Deputy Executive Officer of the Energy Institute at Johannes Kepler University. Furthermore, he is a part of the coordination team of the Austrian power-to-gas platform as well as the Hydrogen Initiative. His methodological research focuses on economic, techno-economic, and systemic analysis in the energy sector, as well as the comprehensive assessment of technologies, systems, measures, and market developments. The content of his research focuses on energy storage, energy networks, and power-to-gas/hydrogen.

Sebestyén Tihamér is Ph.D. Researcher at Green Energy Innovative Biomass Cluster; he completed his studies of territorial planning and geography at Babeş-Bolyai University in Romania. He has five years of experience in researching a diversity of subjects in the field of bioresource and renewable energy, such as biomass and forest residuals. He consolidated his research expertise by gaining practical experience in efficient biomass utilization for energy and awareness raising through several research projects: CT Fellowship, Hungary, and Forerunner Fellowship, Norfolk, USA. At the moment, he is involved in multiple Horizon 2020 projects, such as: PHOENIX, BioVill, BioEnergyTrain, and AgroBioHeat.

Chapter 1
Introduction

Yoram Krozer and Michael Narodoslawsky

Bioresources, all goods of biological origin, are used in the production of energy, food, paper, clothes, in construction and for other products, as well as for leisure, tourism, education, science, arts and other services. They are based on the primary production of biomass, which includes harvested and processed algae, fungi, plants, crops, wood and plant residues, and in secondary production using animals and animal residues. Bioresources are considered non-exhaustible because after the disposal of products based on them, they close material cycles and disintegrate into nutrients for plants, animals and ecosystems. In addition, bioresources provide a sink for pollutants. The Economics of Bioresources is concerned with the value of bioresources. This book is focused on opportunities in bioresources. The possibilities of valuable uses alongside fostering nature are presented.

1.1 Opportunities

Bioresources have unique, beneficial properties in economies. One unique property of bioresources is that the end of a product's life automatically closes material cycles because it provides materials for new resources. This property perpetuates a flow of goods accessible to all plant and animal organisms, including people, referred to as non-exhaustible resources. Another unique property is that they break down in the environment into consumable carbon dioxide (CO_2), water and minerals, referred to as being 'degradable'. The degradability of bioresources differs in terms of the persistence of mineral-based materials that dissipate as pollution and ultimately enter

Y. Krozer (✉)
University of Twente - CSTM, Enschede and Sustainable Innovations Academy, Amsterdam, The Netherlands
e-mail: krozer@xs4all.nl

M. Narodoslawsky
Institute of Process and Particle Engineering, Graz University of Technology, Graz, Austria

Y. Krozer and M. Narodoslawsky (eds.), *Economics of Bioresources*,
https://doi.org/10.1007/978-3-030-14618-4_1

into the food chains of organisms; a recent example is the dissipation of plastics into microscopic compounds that disperse in water and accumulate in food called 'plastic soup'. The third unique property is referred to as their 'regenerative role'; this is their ability to accumulate pollution and neutralize noxious compounds because organisms convert them into materials useful for life. This regenerative capability is essential to the absorption, storage and degradation of noxious compounds, including the storage of carbon dioxide. Such unique properties should be fostered and enhanced.

In economies, bioresources are produced by nature and cultivation; then, processed in various industries and used in many services. Finally, they are consumed and disposed of in a manner that entails degradation and consumption by other organisms which closes material cycles and finally generates new bioresources. Such a bioresource-based economy can, in theory, generate nearly all goods for consumption; thereby creating value while (temporarily) absorbing pollutants. In practice, green plants globally fix about 105 Gt of carbon (1 Gt = 10^9 tons) per year as Net Primary Production (NPP), while economies appropriate 23.8% of all terrestrial Net Primary Production (Beer et al. 2010; Haberl et al. 2007). They provide a sufficient material basis for satisfying the consumption of all people.

The impressive biomass production is realized thanks to the progress of technology in the development and use of mineral fertilizers for soil enrichment; chemicals for pest control; as well as advanced equipment in agriculture and industries. This meets the needs of all 7 billion people in the world. The negative impacts of those technologies are the high use of water and energy; overextraction of exhaustible, mineral resources for fertilizers, pesticides and other chemicals; associated pollution; loss of jobs on farms; and, degradation of nature and environmental qualities. Liabilities for these impacts, however, rarely exist. A lot of bioresources are also wasted as their loss and disposal is often cheaper than sound technologies. Therefore, in this century alone, more sustainable innovations are required in order to prevent those impacts, and satisfy the demands of 11 billion people. The goal of this book is to underpin several opportunities in doing so.

Although the bioresources grew from 12 billion to 19 billion tons a year from 1975 to 2005, its share in materials consumption declined from 38 to 31% because the consumption of minerals for construction and fossil resources was higher (Krausmann et al. 2009). In 2015, the bioresources mass per person was about 3200 kg per year, a number that grew by an annual average of 1.5% during the three preceding decades when compared to a lower population growth of 1.3%. Therefore the idea that bioresources are insufficient for all 7 billion people or more needs correction. A more accurate perception would be that bioresources are abundant, and increasingly available to people who can pay for them, while many others have limited access or cannot afford bioresources. Moreover, producing, processing, and consuming of bioresources by those who can afford them can be improved because those activities are often wasteful; wasteful means that useful materials are disposed rather than reused as secondary products.

The growing mass is not reflected in the higher value of production. On the contrary, all those bioresources produced by agriculture, horticulture, husbandry, forestry and aquaculture generated a global income of USD 2429 billion in 1995 and

USD 2834 billion in 2015. That production value grew at an annual average of 1.1% in the current value of the U.S. dollar, but when corrected for the consumption prices it declined annually by –3.2% in the constant value of USD in 2005. World Bank data also shows that the total global production value has grown in that period of time, from USD 23,927 billion to USD 75,845 billion, a 5.2% nominal growth or 1.9% real growth in the constant value of USD in 2005. Hence, the share of cultivation in the global economy decreased from 7.8% in 1995 to 3.9% in 2015; the share of industries from 33.8% to 27.1%; whilst the share of all services increased from 58.4% to 69.1%; and, within all services, the share of knowledge-based services grew. Opportunities for income generation are primarily found in more knowledge.

Many innovative activities in bioresources emerged with the aim of generating higher value; a few examples elucidate the activities in Europe. Capital-intensive horticulture in Dutch greenhouses, for instance, is high value and low polluting if managed as a closed loop and with energy-efficiency. A second example being localized forestry in the Romanian villages that deliver valuable wooden products and foods along with the production of biofuels for villagers. Ecological agriculture flourishes in many Italian regions where food and tourism are combined based on rich biodiversity. To further exemplify, urban gardening in several French towns delivers meals and leisure, fostering urban ecosystems and public health. In Germany, natural fibers from agriculture generate high value products for industry and construction. Bioresources can generate products of high value and contribute to diversity of nature.

Policies are preoccupied with the scarcity of bioresources, though cultivation delivers cheap bioresources—in bulk—to all kinds of industries. The scarcity argument is often used as a justification for policy support to intensive cultivation. However, bulky deliveries do not provide high product value as novel rival products emerge in the market, with increasing frequency. Policy support for cultivation can turn towards higher value products when they focus on diversification and biodiversity tuned to regional endowments, rather than cultivation of bulk materials. If the present policy support of nearly €63 billion euro a year in the European Union[1] is shifted toward the tuning of bioresources to local ecosystems, the primary products may reduce by volume but their value can grow. The secondary products increase and form an additional resource base. Such shifts are in line with the growing demand for knowledge-intensive, higher quality products and services.

1.2 Innovative Shifts

Biofuels, which are energy products based on bioresources, get much attention in businesses and policy making because many consider them as an interesting option for cultivation and a viable replacement of fossil fuels that emit CO_2 and other pollutants. There are also concerns about the large scale cultivation for biofuels as

[1] https://ec.europa.eu/agriculture/statistics/factsheets_en (accessed 8-1-2018).

this can compete with food cultivation and degrade nature while policy support to biofuels can distort prices. Alternatives can be more attractive.

Additional large scale cultivation for biofuels, the so called 1st generation biofuels, faces changes in energy markets. These changes impede growth toward its mass use. Data from the International Energy Agency shows that about 130 million ton oil equivalent was consumed in 2015, equaling PJ 5451 or 1514 billion kWh. This is a large volume of energy resources but its growth has slowed down during the last few decades. While biofuels cover about 10% of the energy consumption across all countries, it is only 6% in high-income countries where more biofuel can be consumed but the total energy consumption declines. On the contrary, it is found that in low-income countries where biofuels cover more than 80% of all energy consumption, their growth saturates while the total energy consumption increases. In addition, the value of biofuels is usually lower than that of food, feedstock, fiber, industrial and personal products, implying an even lower income to farmers when they turn to cultivating for biofuels. Biofuels are feasible without policy support when much land is available, transport is short and processing cheap; it means that the cultivation of bioresources must be aligned with the local demand and resources. A few cases in the book illustrate this.

The processing of residues from bioresources to biofuels is possible, called 2nd generation biofuels. The scale of residues is large because the disposal of organic waste is cheap, and liabilities for pollution often negligible. Although disposal is often motivated by the enrichment of soil with nutrients, intensive cultivation causes excessive nutrients and mineral fertilizers which pollute the soil, groundwater and surface water. Hence, the disposal of residues is cheap for individual businesses and consumers, but costly to societies as a whole because they must bear the costs of environmental degradation. Biofuels emerge as alternatives to that disposal. The processing of organic waste into biofuels generates energy resources that replace fossil fuels, effectively mitigating disposal, and reduce environmental degradation. Also, the recycling of paper, cloth, wood and other bioresources in the non-organic waste fraction becomes easier when organic residues are collected separately from non-organic waste fractions. When that organic fraction is collected separately, waste plastic, glass and metal can be better separated for recycling. A few case studies address the possibility of the economic use of those residues for biofuels.

The cultivation of algae, fungi and other plants for high value products and services along with biofuels as side-products of processing these bioresources is referred to as 3rd generation biofuels. In refineries based on bioresources, high value specialties can be delivered on the large scale along with biofuels. Small scale cultivation for leisure, education and other services also generates biofuels on the side; for instance, plant residues in parks for local uses. When compared to their rival products based on fossil resources and minerals, the higher prices of products and services based on bioresources form the main barrier for growth because technology and knowledge are less advanced. In addition, the liabilities for pollution and degradation of nature are absent; for instance, liability for CO_2 emissions that cause climate change. In a few areas, advances in the processing of bioresources are tangible; for instance, in bioplastics and tourism. A few case studies illustrate those advances.

1.3 Contents of the Book

The Economics of Bioresources is an introduction to the subject, and is based on theories in economics and engineering exemplified by case studies from various European countries. It aims to motivate students, scholars, managers and policy-makers to pursue further education, research, experimentation and policies in this field. The book begins with a broad overview that narrows down to focus on specific valuable uses of bioresources. This volume briefly touches upon the production of bioresources in agriculture, while focusing on the theories and practices in its processing and uses in industry and service; the agricultural production is covered in other volumes (e.g. Langeveld et al. 2010).

Part I covers theories in 5 chapters. Chapter 2 introduces a few concepts in economic thinking and policies about bioresources: the notion of economic circulation, distribution, cycles and growth. The dynamic nature of economic systems and links to bioresources are pinpointed. Chapter 3 addresses decision making in markets: resources and values, issues in production and transactions, investment decisions and valuation of nature. Issues related to private and common goods of bioresources are highlighted. Chapter 4 is about business development in bioresources: strategies for profitability, development, and business models. Particular attention is paid to innovating in a vested business and starting of a new one. Chapter 5 is about engineering. It begins with the categorization of primary, secondary and tertiary bioresources, followed by a review of the utilization of technologies in production, storage and conversions. Chapter 6 addresses the distribution of bioresources, and in particular adaptations to regional conditions, optimal scale of logistics with bulky transport and planning of a de-centralized system of producing bioresources.

Part II covers cases of bioenergy in four chapters. Chapter 7 is a case study that introduces a production chain which begins with the cultivation of willow right through its processing down to heat production and local consumption in a region of Romania. Chapter 8 is a case study that addresses citizens' interests, demands and participation in the development of district heating in several villages in South East Europe; it also looks into the development of a pilot. Chapter 9 is a case study concerning the integration of bioresources and synthetic gas based on renewable energy, which indicates that the renewable energy can cover a large part of the Austrian consumption at reasonable costs. Chapter 10 is a case study of the processing of manure in the Netherlands aiming at closed material cycles with biofuels and minerals for agriculture as the main products; it presents conditions for beneficial processing.

Part III addresses sectors in three chapters. Chapter 11 indicates the possibilities for sustainable production of bioresources in agriculture. It presents innovative opportunities in conventional businesses and alternative ones. Chapter 12 is a review of the industrial possibilities of high value products based on lignin, which is presently used for biofuel. A range of valuable alternatives in research and development are mentioned. Chapter 13 presents possibilities for the use of bioresource in services, particularly in health, tourism and education which enable the generation of income based on biodiversity.

Chapter 14 summarizes the conclusions of all chapters.

References

Beer C, Reichstein M, Tomelleri E, Ciais P, Jung M, Carvalhais N, Rödenbeck C, Arain MA, Baldocchi D, Bonan GB, Bondeau A, Cescatti A, Lasslop G, Lindroth A, Lomas M, Luyssaert S, Margolis H, Oleson KW, Roupsard O, Veenendaal E, Viovy N, Williams C, Woodward FI, Papale D (2010) Terrestrial gross carbon dioxide uptake: global distribution and covariation with climate. Science 329(5993):834–838

Haberl H, Erb KH, Krausmann F, Gaube V, Bondeau A, Plutzar Ch, Gingrich S, Lucht W, Fischer-Kowalski M (2007) Quantifying and mapping the human appropriation of net primary production in earth's terrestrial ecosystems. PNAS 104(31):12942–12947

Krausmann F, Gingrich S, Eisenmenger N, Erb K-H, Haberl H, Fischer-Kowalski M (2009) Growth in global materials use, GDP and population during the twentieth century. Ecol Econ 68:2696–2705

Langeveld H, Sanders J, Meeusen M (2010) The bio-based economy. Earthscan, London

Part I
Theories

This part briefly introduces several key concepts in the economics of bioresources in order to provide a better understanding regarding the thoughts of economists and engineers about the use of bioresources. First, the notions of circulation, distribution and dynamics of economic processes are introduced with particular attention paid to natural resources, of which bioresources are a part, and possibility of economic development that does not preclude healthy nature. Then, decision making about natural resources is briefly covered including decision support tools, as well as deficiencies in those decisions and possibilities for improvements with respect to bioresources production and consumption. This is followed by a presentation of business strategies for profitable investments and unlimited opportunities for innovative business models focused on bioresources; diversity in entrepreneurship is underlined rather than the promotion of an ideal entrepreneurship. Thereafter, bioresource economics is presented from the perspective of engineering. First of all, the diversity of bioresources with respect to their material characteristics is introduced because these determine production processes, delivery of products and their consumption. Next, technologies for the processing of bioresources into fuels are covered. Finally, the concept of an integrated biorefinery which aims to deliver various industrial products is introduced. Such biorefineries rely on optimized distribution systems and planning of the regional development of bioresources.

Chapter 2
Economic Systems

Yoram Krozer

Economics is a normative knowledge that aims at the creation of welfare with scarce resources. An economic system covers interactions between production, consumption, and their institutions, within a country, region, or community. Throughout the last century, the income of global economic systems has quadrupled after price correction, though achieved at the cost of social conflicts and degradation of natural qualities. Rarely is nature at the core of economic thinking though it is essential to welfare. Herewith, economic circulation and distributions are introduced, followed by presentations of economic cycles and growth, all with particular attention to nature and bioresources.

2.1 Economic Circulation

'Economics' combines the Greek words *oikos* (home, property) and *nomus* (managing) to mean knowledge about the household. Its normative goal is a contribution to welfare to be accomplished by—satisfying individual and social demands and aspirations; the distribution of wealth, leisure, care, and other values; enabling decision making about these values across generations, sexes, races, and so on (based on Sen 2009). Toward achieving this goal, nature, labor, capital, and knowledge are combined in the production of goods for consumption usually measured in money, although many valuable goods are not monetized, for instance the diversity of living organisms.

Natural resources are considered to include all minerals, fuels, organisms, as well as all environmental qualities. Herewith, bioresources refer to the organisms, in the ascending order of complexity starting from genes, enzymes, cells, viruses, bacteria

Y. Krozer (✉)
University of Twente - CSTM, Enschede and Sustainable Innovations Academy, Amsterdam, The Netherlands
e-mail: krozer@xs4all.nl

Y. Krozer and M. Narodoslawsky (eds.), *Economics of Bioresources*,
https://doi.org/10.1007/978-3-030-14618-4_2

to plants and animals, as well as all the compositions or organisms referred to as ecosystems which are compounded into landscapes and finally, into biomes. Labor refers to all activities of people, ranging from production in agriculture and processing in industries—which are usually material-intensive activities, down to the material-extensive leisure, education, finance, arts, and so on. Capital encompasses all tools for production and consumption which cover material-intensive constructions and equipment, as well as the material-extensive artifacts of savings, cash, and other valuables. Knowledge addresses the comprehension of interactions between these resources, pertaining to economics and our capability to transform scarce resources into valuable products referred to as 'know-how'.

Economy is comprehended as a system in which resources are combined into valuable goods that circulate from production to consumption, whereas awards for those goods—expressed in money or any other symbol of value—circulate in the opposite direction. This circulation involves exchanges in which consumers deliver labor to producers at a cost, and purchase products at a price, while producers deliver income to labor and supply goods for the purchases. For the sake of convenience, the supplies of all producers are called products whether they are material-intensive or material-extensive ones.

2.1.1 Accounts of the Circulation

The global economy is considered a closed economic system but countries, regions, and communities are open economies because some resources are purchased from abroad, i.e., imported, and products are sold abroad, i.e., exported. Such circulation expressed in money is shown in Fig. 2.1, which illustrates the summed up value in money, referred to as the 'aggregated product', as well as the flow of money within an economic system, called 'transfers'.[1]

In a closed economic system, all expenditures equal all income and to all product supplies in money; the latter refers to Gross Domestic Product (GDP). In an open economy, exports and imports are added and subtracted with '+' and '−' signs, respectively, to obtain the Gross National Product (GNP). Exports plus imports is referred to as 'trade', whereas exports minus imports are called the 'trade balance'. Small economies usually have a larger share of trade in the Gross National Product than large ones, because they possess less local resources but international trade is also constrained by 'barriers of trade' which are policies aimed to enhance the domestic economy.

The Gross National Product equals the Gross National Expenditure and Gross National Income. This is valid for any household at a moment in time be it household of a nation, region, or municipalities. Note that a community or region can be larger than a country measured by the aggregated income, e.g., the economy of California is larger than many countries in Europe and the largest cities in Europe are larger

[1] More explanation is on YouTube (3 parts): https://www.youtube.com/watch?v=ail9Jr9UjiM.

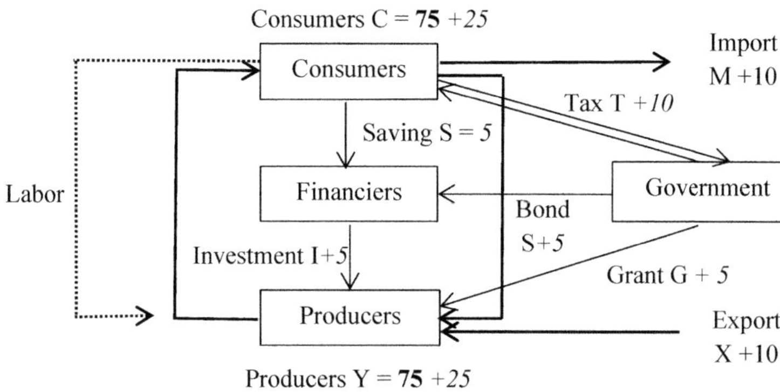

Fig. 2.1 A simplified schematic economic circulation with fictive numbers; in bold are the domestic aggregates, in italics are transfers within the system

than the economies of Cyprus or Malta. The aggregate values are in balance at any point of time as shown in Box 2.1.

Box 2.1 Definitions in the economic systems

$$\mathrm{GNP} = \mathrm{GNE} = \mathrm{GNI}$$
$$\mathrm{GNP} = \mathrm{GDP} + X - M$$

When GNP is Gross National Product, GNE is Gross National Expenditure, GNI is Gross National Income, GDP is Gross Domestic Product; M is Import, X is export but various letters are used.

The economic systems also involve circulations within the aggregated consumers' expenditures and producers' supplies. The consumers' incomes are spent on personal consumption, taxes for authorities, and savings in financial institutions called 'private consumption'. The taxes are spent by authorities for distribution across consumers and producers and on investments, called 'public consumption'. The consumers' savings are used by financial institutions for investments, loans, and other capital resources; they are also used to pay taxes. These transfers imply that—the aggregated income is equal to the aggregated consumption plus taxes and savings in the financial institutions; the taxes equal to the income plus expenditures of the authorities; and that all savings mirror all investments. See Box 2.2.

Box 2.2 Definitions in distribution

$$\text{GNI} = Y = C + S + T$$
$$S = I$$

Y is Gross National Income, *C* is consumption, *S* is saving, *T* is taxes, and *I* is investments but various letters are used.

The division between private and public transfers is fluid. The authorities also operate as financial institutions with monopoly powers to supply money and provide 'bonds', which are loans with guarantee payment. Some private financial institutions do not use the money printed by authorities but operate due to credit and surrogates as Bitcoin and when the financial institutions are 'too large to fall' they behave as if an authority; note that some financial institutions are larger than the economies of many countries.

The monetary transfers do not indicate transfers of energy and materials because values of natural resources differ; for instance, the average price of gold per mass unit is thousands of times higher than that of wood though more wood mass is consumed than gold. High value of materials shows that people appreciate qualities; the appreciations are determined by functional qualities, consumptive patterns, image, conventions, efforts in extraction, and other factors. However, material flows are not indicative of monetary flows and the values of products should not be confused with their mass.

2.1.2 Assessing Welfare

Measuring contributions of activities to welfare is a challenge. In engineering, combinations of resources that deliver valuable products are measured in mass, energy, space, and other physical terms; which can be tons of grain per hectare of land, or the consumption of biofuels in energy units per hour in kilowatt hours (kWh). The useful applications of materials and energy as outputs compared to materials and energy use as inputs are referred to as the 'material efficiency' and 'energy efficiency', respectively. The economic yardstick is value measured by price. Hence, the GDP per material consumption or energy consumption, referred to as 'material intensity' and 'energy intensity', indicates the value of their uses. The GDP per cost of research and development (R&D) or per number of scientists is referred to as 'knowledge-intensity' used to assess innovativeness of economies. The GDP per labor hours is referred to as 'labor productivity' and GDP per capital stock in prices as 'capital productivity'. These are main indicators of economic performance and conventional benchmarks for wages, capital performance, income growth, and other interests. The total factor of productivity is outputs including inputs that cannot be explained by labor and capital productivity, which usually refers to technological change.

Since people pursue a higher income—people are particularly anxious about losing it—policies aim at a larger aggregated income measured by the GDP and to prevent a decline of GDP, called recession. The GDP, however, does not disclose if people are educated, water is drinkable, biodiversity is rich, and other goods that contribute to welfare are in place. GDP grows whenever an aggregated expenditure increases, even if more pollution is generated and more spent on pollution controls. Expenditures can counter such negative impacts of activities but cause higher prices because money is spent without the sales of valuable products. The GDP corrected for the prices of typical purchases, which shows purchasing power parity (PPP), is a better indicator of welfare but such Net Domestic Product (NDP) is not often used in policymaking. Many economists argue for several indicators because many values are not priced, for instance, health, happiness, and nature. A broader set of the welfare indicators evolved throughout the last century as shown in Table 2.1.

Income is important for welfare but it is not the only indicator. Education and health care contribute to increased life span and enhances the quality of life. The Human Development Index (HDI) combines the life expectancy at birth, years of schooling for adults above 25 years and a logarithm of income per capita into a dimension-less number for comparison between countries. The North European countries are on top, the Central African countries at the bottom. The Happiness Indexes indicate welfare as expressed in opinions and statistical data about quality of life with respect to social security, community interactions, personal opportunities, and other factors of individual well-being. Similar countries can be found on the top and bottom. Environmental economists pursued indicators for natural resources and environmental qualities, such as the stock of bioresources and biodiversity, pollution and pollution controls, and others. These environmental indicators are used widely nowadays. More indicators are suggested, for instance, ownership of land and capital because these enable entrepreneurial activities and security of income by a guaranteed basic income.

Table 2.1 Indicators of welfare in the last century

Demands for	Value of	Indicators are
Production (1920s on)	Income	Gross Domestic Product (GDP)
Consumption (1950s on)	Purchases	GDP-PPP
Capabilities (1970s on)	Education, Health	Human Development Index (HDI)
Satisfaction (1980s on)	Opinions	Happiness Index
Environment (1990s on)	Nature	Impact Indicators

2.1.3 Accounting for Nature

In mainstream economics, nature is exploitable as any other resource. A stock of natural resources is considered as capital created by human activities. Incomes from that capital are also referred to as rents. The rents due to exploitation of natural resources are valued at average commodity prices in international trade. The rents fluctuated between 2% and 6% of the global GDP throughout the last 50 years, as estimated by the World Bank.[2] This is based on the stocks of main natural resources: oil, gas, coal, and a few key mineral resources as well as forests that are a bioresource. The stock value of the bioresources covered about 12% of all those rents, which means less than 1% of all global income. Although the stocks of natural resources are important for some economies due to large reserves of oil, gas, forests, or other natural resources, income from labor and capital constitute more than 90% of the GDP in all countries, even more than 96% in the European Union; rents from R&D measured by licenses are even lower than from natural resources. Such rent assessments, however, are disputed because the scale and scarcity of stocks are insufficiently accounted for, and because losses of natural resources, degradation of environmental qualities and ecosystems are neglected.

Mineral resources are concentrated in some places on Earth where they are exploited and transported to places for their use; they are thus displaced. After being used, the mineral resources are not transported back to the place of origin but shift to another place. Since the mineral resources are displaced, they are therefore considered as exhaustible, non-renewable resources. Many social and environmental capabilities are also degraded. The displacement and degradation entail costs. However, these costs are not paid unless the exploiters are liable for the degradation, which is rare. After the uses, the mineral resources do not disappear but remain, being disposed nearby to where they are used. When they dissipate, referred to as emissions that cause pollution, they damage human health, nature, and environmental qualities. It is also a cost that is largely unpaid because polluters are rarely liable for their emissions. Some costs are paid from public means. The unpaid costs of the degradation and pollution are referred to as the social costs or environmental costs.

Bioresources are better distributed on earth when compared to mineral resources, though the varieties of bioresources depend on local environmental conditions. The spread implies that similar bioresources can be produced in many places but in tune to local conditions. After their use, the bioresources degrade into organic materials which are nutrients for the creation of new organisms. These organisms become new bioresources. This material cycle evolves in nature at no cost and generates the growth of bioresources, given the environmental conditions. Hence, the bioresources are referred to as the 'non-exhaustible, renewable resources'. The natural degradation of bioresources does not pollute after disposal because most organic materials are broken down and reused in nature after a period of time with the exception of some minerals, e.g., in bones. The natural recirculation and degradation of bioresources

[2]https://data.worldbank.org/indicator/NY.GDP.TOTL.RT.ZS http://wdi.worldbank.org/table/3.14 (last access 11-12-2018).

rather than pollution are major advantages compared to mineral resources. However, these advantages of the bioresources are not reflected in economics as long as the social costs are not expressed in decision making.

The mineral resources and bioresources also involve transport, processing, and consumption that cause scarcities and pollution but the damages are often neglected in economic records, thinking, and planning. The damages are considered as external effects to the economic systems. When the social costs of such external effects are estimated, they usually exceed ten percent of the GDP, even as much as one-third of the GDP in countries whose economies are resource-rich and material-intensive. Corrections of the GDP for the costs of damage, however, are not popular in policy making, possibly because they reduce the GDP rather than show its growth. The scarcity of natural resources and pollution as estimated in energy and material balances and environmental assessments can be compared to income but they gain less attention in decision making than GDP. Damages caused by the external effects can be systematically factored into the cost, income or prices, referred to as the 'internalization of the external effects'.

2.1.4 *Inflation and Purchasing Price Parity*

Income per person does not reflect welfare if the consumers' prices increase. The price increase is called 'inflation'. Various factors can cause inflation. Demand for higher wages cause higher production costs, entailing higher prices. Demand for capital can drive up the cost of capital, for instance demand for housing can drive up the housing prices. High prices of key natural resources, e.g., oil and gas, can invoke inflation. Inflation is also triggered when products are scarcer than money because authorities obstruct production or because they print money when aiming to satisfy public opinion without the balance of available products.

A low rate of inflation, a few percent annually, is easily met in consumer purchases but high rates cause uncertainties about future prices. When higher prices are expected, people claim higher salaries and hoard products which accelerates inflation. If money loses its purchasing power, many people become penniless billionaires as experienced in the last fifty years among those in Poland and former Yugoslavia during the revolution of 1980–81, and before the civil war of 1991–93, respectively. Matching consumer prices to the labor costs in production is a remedy, but it implies a huge loss on the purchasing power as experienced in all former communist countries during the 1990s and part and parcel in all capitalist ones based on authoritarian regimes, whether it is formerly rich Venezuela or poor Zimbabwe.

The price decrease is called 'deflation', typically caused by a sudden loss of income, for instance following fall of the banking system in 2008 in the United States and Europe. Deflation triggers expectations about decreasing prices and people withdraw purchases resulting in a slow-down of economic circulation, in a phenomenon known as 'depression' meaning a long period of economic stagnation and decreasing welfare. Depression is difficult to resolve because low-income impedes consumption;

that said, some strings of environmental thinking welcome deflation because lower consumption is assumed to reduce the negative impacts on environmental qualities.

Inflation and deflation are expressed as a percent price increase or decrease compared to a reference year. Income corrected for the rate of inflation and deflation based on a typical basket of consumption is referred to as 'purchasing price parity (PPP)', 'real income', or the 'income in constant value'. The PPP is related to the money value of a particular year, for instance, euro of 2000, written as €$_{2000}$. When the purchases are realized abroad and sales on the domestic market, or vice versa, the sales are also corrected for the exchange rate of currencies. Statistical offices usually estimate incomes in current prices, also called 'nominal prices'. Some statistical offices correct the current prices for inflation and convert them to a uniform currency, which delivers PPP data based on a particular year as shown in Box 2.3.

Box 2.3 Definitions of inflation and deflation

$$a = p_{t+1} / p \cdot 100\%$$

For a is inflation (+) or deflations (−), p is consumer prices and the real income Y_r is

$$Y_r = Y_t \cdot (1 - a)$$

And the real income growth g_r is:

$$g_r = Y_{r+1} / Y_r$$

For example, if the gross income grew 5% in the first year and 2% in the second year, the annual average gross income grew 3.5%; but, if the annual inflation was 3%, the Net National Income grew 2% and −1%, respectively, which is 1% in total. The economic slump in the second year is hidden by the gross growth.

The currency conversion is:

$$b = V_b / V_a$$

When b is exchange rate and V is the country's currency.

Though inflation and currency often go hand in hand, a particular currency can appreciate or lose value. For instance, inflation in US dollars (USD) in 2015 was 1.3% in the United States compared to 1.2% in the European Union, the inflation in the United States was somewhat higher, but the value of the Euro declined from 1.3 USD in 2015 to 1.1 USD years later. The exchange rate decreases when domestic money depreciates. For an international comparison, income is corrected for inflation and currency based on a standard—that standard was gold in the past, and is presently the US dollar. A lower value of currency, devaluation, implies costlier purchases abroad but cheaper export of domestic production because it is priced in the domestic currency.

Policies aiming to boost exports devalue their currency. It works if the devaluation is incidental and importing countries do not devaluate as well. If devaluation persists, many people lose trust in their economy and purchase foreign currency. If the mistrust continues people may even massively buy foreign currency causing accelerated devaluation and loss of purchasing power because the imported goods are unaffordable when people are paid in the local currency. That local currency loses value and people use the standard currency, as experienced in many authoritarian countries.

2.2 Economic Distribution

Economic distribution is at the core of economic deliberations but briefly touched. It has many dimensions: international transfers, division in production and consumption, disparities in wealth and income between countries and regions and between people within a country, as well as the division of rents and wages. The distribution of ownership and welfare can also be considered but addressed in Chap. 3.

2.2.1 International Transfer

Imports and exports comprise international transfers. These transfers are costly and risky, but pursued when the exporting suppliers expect more benefits abroad than at home, and when the importing demanders expect to accrue cheaper or better production compared to domestic supplies. However, the transfers are contentious because the local natural resources are usually promoted even when their prices or qualities are inferior compared to the foreign ones. Local bioresources, herewith, are often valuable due to the specific taste, smell and other characteristics. Imports of products are also obstructed by the preferential treatment of domestic suppliers, even if the imported ones are cheaper or of better quality. Attributes of domestic labor, capital, and knowledge are generally prioritized as well, in particular when related to local crafts and traditions.

Transfers of labor, called migration, is often opposed by inhabitants even if migrants add skills to the local economy and execute tasks perceived as inferior by others. Transfers of capital are flexible when money is involved, but disputed when domestic rules are evaded and trade with capital embedded in equipment is often restricted by certifications. The knowledge transfers embedded in equipment are considered as capital, referred to as ‘Foreign Direct Investments’. Knowledge disembedded in skills and crafts, called ‘know-how’, is considered as labor referred to as ‘technology transfer’. Such knowledge transfers are opposed if it is assumed that they will magnify the dependency on foreign ownership or patents, even though these enhance the capabilities of the recipients.

The idea that trade makes exporters and importers better off if each one is focused on its specialization is a cornerstone of economic teaching, referred to as 'comparative advantages'. Although mutual benefits of the comparative advantages are widely acknowledged, various restrictions are imposed by policies. There are duties on imports and subsidies for exports, tax exemptions, and subsidies offered to domestic producers, as well as standards for product qualities, environmental, and safety performance and other non-monetary barriers of trade. Policies that create barriers of trade for the protection of domestic industries invoke retaliations by other policies resulting in international conflict, even trade wars that often evolve into military confrontations.

Trade is also contentious in environmentalist thinking. From that perspective, it is argued that international transfers undermine high social and environmental standards because they enhance migration of industries to unregulated areas, delivery of substandard products, low environmental performance, illegal capture of land, poaching species, and other malicious activities due to unregulated locations called 'regulatory havens'. Such a 'race to the bottom' is objected to, but is not necessary when stricter standards are pursued. This viewpoint is supported by observations that environmental innovations released in domestic markets due to stringent policies can be exported resulting in better quality products (Copeland and Taylor 2003). Herewith, the frontrunners in environmental performance usually pursue high standards in trade that are opposed by laggards.

Since bioresources depend on the local environmental conditions, the assumption made about comparative advantages implies specializations based on regional capabilities. Bioresources are sometimes considered a possibility for self-sufficiency. The argumentation is that bioresources enable self-sufficient communities because proper land uses can provide all products that meet basic needs in a sustainable manner. Such autarkic economies, meaning nearly closed economic systems, keep economic circulation within the community. They can operate under exceptional conditions of war and famine, the dictated demands by ideologies and religions, or due to temporary exceptional crafts. However, self-sufficient economies on a voluntary basis rarely sustain across generations because they are vulnerable to social and technological changes.

2.2.2 Economic Structure

The distribution of production refers to the agricultural, industrial, and service sectors organized into branches called an 'economic structure', as well as to consumption divided into categories of purchases, called 'consumption patterns'. The global economic structure and consumption pattern has changed throughout the last century. The share of agricultural income in the global income has decreased from one-third to just a few percent; as the share of industrial income has decreased from half to a quarter, whereas services grew to dominate from about quarter to nearly 70% of the global income. These changes are also reflected in consumption patterns because of the growing consumption of services.

The economic structure is expressed in input–output tables, which represent the value of suppliers in rows and demanders in columns. The national and regional input–output tables usually cover a few dozen branches, first in the primary sectors of mining and agriculture, followed by the basic and processing industries (in the service sector) and ultimately the consumers. The sequence of branches reflects the increasing share of labor, capital, and knowledge. Trade is also included. Some tables cover the technologies for pollution controls but stocks of natural resources, scarcities, and damages caused by pollution and waste are generally excluded. See Box 2.4.

Box 2.4 A simple input–output table of economy

Assume 100 value units of consumption. The farmers supply 5 biofuel units to energy producers and 15 to wood-working; the energy producers supply 10 units to the farmers and 10 to the wood-workers; and the wood-workers supply 30 units to the farmers, 25 to the energy producers and use 5 units themselves.

An example of input–output

Suppliers'/Demanders'	Farmer	Energy	Wood-workers
Farmer	0	5	15
Energy	10	0	10
Wood-working	30	25	5

The input–output tables are used to assess the consequences of changes in economies, in particular the distribution of values.[3] A decomposition of the values into resources would show that the share of natural resources decreases as activities move from agriculture and industry to services. The distribution of natural resources across sectors is shown in the material and energy balances, which represent flows in mass and energy units throughout an economic system but its linking to the distribution of values is still deficient in such metabolism studies. The metabolism studies are increasingly pursued because enable to assess progress in welfare taking into account all natural qualities.

[3]Basics on the input–output model: https://www.youtube.com/watch?v=U0hGtZqUbSM; an example of calculating a solution: https://www.youtube.com/watch?v=6C1MOD_oKwU; more detailed explanations in a serial https://www.youtube.com/watch?v=G02IRkwmxXw.

2.2.3 Wealth and Income

Credit Suisse in its Global Wealth Report has estimated global wealth in 2010. The total value of stock in buildings, shares, savings, cash, and other assets minus debts and other liabilities is assessed at 200,000 billion USD, nearly three times larger than the global income of 73,000 billion USD. Wealth grows faster than income because the rents of assets grow faster than that of liabilities. However, it is unevenly distributed. In 2010, less than 1% of people who possessed more than USD 1 million, mostly men, held about 36% of the global wealth, found mainly in North America, Europe, and Asia Pacific. Wealth grows rapidly. The wealthiest ones become wealthier due to rents. Rents from capital, herewith, grow faster than income from wages because capital is more mobile and less regulated than labor as observed across countries throughout the last century (Piketty 2014). This process evolves particularly fast in Mexico and the United States where between the years 1995 and 2014, the share of wages in the Gross National Income decreased from 41 to 35% and 60 to 57%, respectively (ILO 2017).

The main assets are buildings that grow fast concentrated in the urban areas, particularly the city centers. The less wealthy are driven out by higher prices as the wealthier move in, referred to as 'gentrification'; this is encouraged by subsidies, tax exemptions, and other policies in support of construction. When real estate prices grow buildings spread toward cheaper locations which drive mobility. This urban sprawl requires costly infrastructure, captures space, and degrades nature. Wealth concentrated in the cities attracts migrants from rural areas which causes fewer services for the rural population. Bioresources, measured by owned land, are not a major wealth category except in low-wealth countries in which the value of land does not grow that fast. Abundant biodiversity provides opportunities for services based on natural resources, for instance eco-tourism.

The income distribution is measured as an income group's share of real income in the total real income, divided by the share of that income group in the total population, called 'Gini coefficient'; though criticized because does not disclose wealth, division between wage and capital, nor the highest few percent of income. The Gini coefficient is a number between 0 and 1 which means completely equal and unequal, respectively. The global Gini coefficient is above 0.61, but is around 0.4 in most high income countries' per capita (in the OECD); it is even lower in Japan (0.25). The income disparity in the lower income countries is larger than in the high income countries. The national income of countries grows toward each other because people in the low-income countries become richer. Effectively, the income of people in low-income and middle-income countries increased in the span of the past twenty years but did not increase in the high income countries, except in the highest income group. Globalization contributed to a higher income for many poor people in all countries, but not to those who fall in the lower- and middle-income categories in the rich countries (Milanovic 2012).

Consumption patterns become less material-intensive and polluting, but the share of bioresources decreases. More products based on mineral resources are supplied even though the bioresources-based products are appreciated more when opinions are assessed, because these products are associated with nature and a good environment. The growing income generate more demand for the attributes of nature and environment in products, such as healthy foods, renewable fuels, and ecological tourism. Local bioresources, herewith, could be perceived attractive. There is reason to be optimistic about the bioresources in consumption but a clearer understanding of consumption patterns is necessary, particularly in relation to bioresources.

2.2.4 Natural Resources

Natural resources are often neglected in the economic thinking about distribution because they constitute a small fraction of the countries' GDP. Nevertheless, many economies largely depend on the exploitation of a particular resource, for instance of oil or forest reserves. The mainstream viewpoint about distribution is that the natural resources belong to explorers or exploiters; they are considered as private possessions. The alternative, heterodox train of thought is that all natural resources are owned by the communities vested in areas where the resources are located. In this approach, the community rights of natural resources prevail above private ownership.

Possession of natural resources is an asset that can generate high income but the owners do not have much interest in the development of other businesses. Resource exploitation grows, whereas other activities stagnate. In effect, the income and sectorial disparities in the economy increase. Also, when policy is preoccupied with resource exploitation, novel activities are considered less relevant and so sectorial base of economy remains narrow. The governmental involvement often deepens the disparities because well-earning businesses are supported entailing corruption, as well as health risks and environmental damages when these interests are protected. Even if the incomes of a few businesses grow, welfare declines—labeled as the 'curse of natural resources' (Sachs and Warner 2001). The spatial concentration of exploitation generates islands of prosperity among poor regions leading to accelerated migration, growing social tensions, political conflict between cultural groups, and sometimes even war.

Several options can be found in the community-based approach. A market-based option is sharing income from the exploitation among all community members. For example, all adult inhabitants of Alaska in the United States receive an annual payment based on the exploitation of oil resources in that state. Another market-based option is the establishment of a development company that is entitled to distribute income from the exploitation among innovating firms, an example of this is applied in Norway. The state-based option is participation in the exploitation of natural resources which generates income in addition to taxes, as it is arranged for natural gas in the Netherlands. Another state-based option is taxation of the exploitation or external effects, for instance, space use or carbon dioxide, respectively; the latter is found, for example, in Sweden. The ownership-based based approach generates high income but the community-based approach usually delivers a fairer distribution of wealth and broader economic basis.

2.3 Economic Cycles

Since the demand for products change, as do the combinations of resources and their costs, productivity fluctuates accordingly. Large productivity fluctuations are driven by seasonal changes, production capacity, new businesses, and new technologies. Each one is briefly covered hereunder.

2.3.1 Fluctuations

Seasonal fluctuations in volume are common in all products. Particularly fluctuating are bulky products based on bioresources, for instance, grain. The prices of such commodities fluctuate even more than their volumes do and are hence called 'volatile markets'. The prices go down when bioresources are abundant during high season, for instance at harvest, and by the same logic, they go up during the low season when bioresources are stored and become scarce. Market volatility decreases when storage capacity is high, harvesting goes on throughout the year or imports complement domestic commodities. Further, volatility increases when storage deteriorates, transport for import is costly or during social conflicts because during such times, scarcities cannot be resolved instantly until after the harvests. Like this, seasonal cycles continuously evolve within a year.

Changing demand for a product also causes fluctuations because increasing production to meet them is time-consuming, for instance, the delivery of biofuel from forestry takes a few years before trees can be harvested. When a large producer expects more demand in the future, capacity is expanded to gain a market share. If more producers follow that frontrunner, the expanding production capacity will eventually exceed future demands. The overcapacity causes lower product prices due to abundant production, entailing a breakdown of producers and lowered capacity, followed by scarcer products for any given demand and higher product prices. Such fluctuations, referred to as the 'pork cycles', occur whenever expansion needs time and the products meet the basic demands, for example, housing, food and energy. The cycle of production capacity evolves spontaneously and is difficult to prevent because the predictability of prices, demands, and capacities is imperfect. Fluctuations can be dimmed when alternative suppliers are made available, for instance, bioresources in addition to mineral resources. Production cycles evolve every four to ten years.

Fluctuations also emerge when new production methods and products enter markets creating new businesses, referred to as 'innovations'. Innovations often emerge when resources become unreachable or costly because entrepreneurs must look for alternatives. For instance, scarce land invokes innovations that save the cost of land through more capital, as seen in greenhouses. That said, when much land is available while labor is scarce, then costly machinery for large agricultural holdings is developed and used. Innovation can also add costs but deliver outputs that are larger or superior in quality, for instance, innovation in breeding can generate a seed that enables larger biomass per hectare or provides highly appreciated qualities. Innovations that reduce the costs of production (process innovations) or improve the qualities of products (product innovations) are followed by dissemination across sectors, referred to as 'diffusion'. Innovations enable the entry of new businesses that outperform the vested ones by delivering a lower cost or better quality products, called 'creative destruction' (Schumpeter [1939] 1989). These business cycles emerge every 15–25 years.

The entry of new technologies can catalyze changes in the economic structure as they introduce new production and consumption patterns, such as (in the last century) electricity, agro-chemicals, and informatics. New technology generates businesses in various sectors, referred to as 'spin-offs'. For example, the agro-chemical industries delivered fertilizers and pesticides to agriculture during the 1950s which enhanced the production of biomass per hectare that we now refer to as the 'Green Revolution'; though at a social and environmental cost of pollution, degradation of soil and loss of farm labor. The increase in agricultural productivity generated spin-offs in industries, such as storage technologies and prepared foods. While growth in agricultural productivity may saturate, in the meanwhile, genetic engineering has created high expectations about even higher agricultural productivity though the results are, so far, disappointing. Major technological changes are also observed in energy, as the share of biomass and peat in the global consumption has decreased in favor of coal, oil, and gas. Soon to follow was the entry of modern renewable energy such as modern biofuels, geothermal, wind, and solar power, whereas hydropower and nuclear power retain a low share in the total energy consumption (Grubler 2012). Such technological changes evolve every 25–50 years and are called 'long waves'.

2.3.2 Multiplier

Fluctuations cause a loss in welfare when economies slump. For this reason, many economists advocate additional expenditures that generate a stream of subsequent expenditures referred to as the 'multiplier'. For example, an additional expenditure of 1 euro on bread is passed to a baker, thereafter to a miller, then to a grain trader, and eventually to a farmer. When some expenditure is allocated in savings, taxes to the authorities and are spent on imports; then, total expenditures minus savings, taxes, and imports remain in circulation. Savings and taxes are returned into circulation as private and public investments, but imports leak from that circulation.

The multiplier is usually created by the additional expenditures of governments based on loans that need to be repaid, including rents to the money suppliers after the slump—for instance, loans to boost forestry should be covered by income from forestry when the trees mature. The multiplier fades when investment based on savings and taxes are unproductive, and imports are high; large indirect income due to the multiplier is generated when the investment is productive and leaks are small because resources circulate locally. See Box 2.5.

Box 2.5 A multiplier

$$I_t = a \cdot (1 + I)$$
$$a = C - (M + S + T)$$

When I is income, a is multiplier, C is consumption, M is import, S is saving, and T is tax.

If in an economic system of 1000 units, 50 units of government expenditure is added, it generates 25 units of additional consumption expenditures—if 15 units leak as imports, 5 as taxes and 5 as savings. The multiplier is 25/50 = 0.50, which means 0.5 of the expenditure can be added as indirect income to the direct income, or 1 + 0.5 total income.

The multiplier is at the core of economic policy debates. The issue is whether economic slumps should be countered by an increase in government expenditures as argued by the Keynesian economists, after John Maynard Keynes, or by lowering government expenditures because it reduces debt which generates lower rent for investments as recommended by the economists branded as the 'Monetarists'. The former Keynesian policy successfully countered the depression during 1930 after the financial crash of 1929, and generated welfare states in Europe during the decades after the 2nd World War. The Monetarist perspective dominates in policies from the 1990s onwards and generates income growth alongside a larger income disparity. The economic depression after the financial crash in 2008 was resolved by Monetarist policies at high social costs because millions in high income countries were reduced to poverty and environmental degradation accelerated. Many economists argue that government expenditures would faster resolve that crisis.

The environmental-economic perspective is ambiguous. Environmental activists often advocate supplies-based local resources and circulation of money within community; several communities established local currency for that purpose. Low imports imply high multipliers. This advocacy, however, creates a barrier of trade and results in costlier resources than the traded ones when scaled-up for use in many communities. Policy perspective is also ambiguous. State intervention is considered necessary for the availability of natural resources which finds support in the Keynesian argumentation but low, preferably nil, rent is also advocated which is rooted in the Monetarist view. A more coherent environmental—economic perspective on growth and distribution would add to that debate.

2.4 Economic Growth

Economic growth means higher value of the aggregated product or income for a period of time after the corrections for inflation and fluctuations have taken place. Given economic circulation, the higher value implies a larger demand for products. The demands are assumed infinite and externally given, while they do not necessarily consume more natural resources. On the contrary, reallocation of resources from material-intensive agriculture and industry toward more valued material-extensive services generates income growth, as well as less material consumption and pollution. Material-intensive economic growth can indicate inefficient resource allocations, in particular ineffective use of knowledge for valuable products.

2.4.1 Drivers of Economic Growth

The mainstream economic thinking about economic growth addresses the scale and productivity of capital and labor. The scale of capital refers to investments, and the scale of labor to the working population. The productivity of capital is assumed flexible, thereby nearly perfect, though it is highly concentrated and causes economic crises. Labor productivity is considered as the aggregated product per labor cost. As per this viewpoint, for economic growth, imperfections in allocation of labor should be resolved. Knowledge is comprehended as labor when it involves know-how or as capital when embedded in technology. See Box 2.6.

Box 2.6 The mainstream view on economic growth in the Cobb-Douglas model:

$$Q_{kl} = A \cdot K^{\alpha} \cdot L^{\beta}$$

For a period, Q is the increase in GDP, A is a constant for technological change, K is the capital productivity approximated as value of machines, equipment, and construction, L is the labor productivity being valued per man-hour cost, whereas α and β represent the substitution of capital for labor ($\alpha + \beta = 1$ under perfect competition, α/β is often constant).

As sectors shifted from agriculture to industries and then to services throughout the last century, the global income has grown an average of 1.6–1.8% in real prices compared to 0.4–0.6% average growth of materials measured by mass. In effect, the global economy dematerialized. The dematerialization implies that growing economies usually become less material and energy consuming per unit of income even though their total material consumption can increase when economies grow. Since pollution is usually linked to the use of natural resources, the pollution volume per production unit also decreases as production and consumption of natural resources decreases though the total pollution increases as the scale of production and consumption enlarges. Incomes can grow along with lower material use per income, called 'relative decoupling' or even lower material use in total, called 'absolute decoupling' (Fischer-Kowalski and Swilling 2011). The rate of decoupling is the rate of income growth minus materials growth which varies between −0.2 and + 0.8% but rarely goes above 1%; below 1% indicates a relative decoupling and below 0% indicates absolute decoupling (Dittrich et al. 2012). The idea of income growth along with accelerated decoupling is embraced in policy making as 'green growth'.

An explanation is that cost-saving due to material and energy efficiencies in production is allocated in higher value products, which generates income growth along with lower consumption of energy and material, and pollution reduction. Alternatively, cost-saving can be distributed among low-income groups; it can be focused on jobs in poor communities, used for protection of ecosystems, restoration of environmental qualities, and other purposes.

Environmental qualities rather than income are favored from the environmental perspective (Daly and Cobb 1989). A balance between higher incomes, social gains, and environmental qualities is often advocated and popularized with labels such as 'Triple P' (Profit, People, Planet by Elkington 1997) and 'Donut Economy' (Raworth 2017). Such allocations rarely evolve spontaneously in markets because though they increase welfare, it is assumed that they impede income growth. They are driven by an authority, usually the national government. A forced distribution without economic growth usually causes societal tensions, whereas a balanced income growth, social relations, and natural qualities makes for the more desirable option.

2.4.2 Innovations

It has been observed that the division between natural, labor, and capital resources is largely driven by technological change embodied in equipment and disembodied as know-how. These observations bolster innovation as driving factors for economic growth, but the drivers of innovation are disputed.

In the mainstream view, innovations are driven by higher resource prices; they emerge because prices increase when resources are scarce. Innovations enable substitutions of less scarce resources, though higher prices can also accelerate the exploitation of natural resources without innovations. An opposite argumentation, referred to as the 'evolutionary theory', is that knowledge generates innovations which enable us to reduce prices of resources. Observations confirm that innovations are often associated with expenditures on research and development, but relations to the innovative businesses are obscure. In this argumentation, natural resources are managed if sufficient knowledge is available. In behavioral theory, entrepreneurs in firms and social organizations create opportunities for innovations with regard to deficiencies in the fulfillment of demands. Since degradation of natural resources generates a sense of urgency in society, entrepreneurs who perceive opportunities pursue innovations; the availability of natural resources depends on private and social demands.

The substitutions of vested technologies for new ones are referred to as 'radical innovations'. These entail numerous improvements in time, referred to as 'incremental innovations', which reduce high costs and deficient performance of the radical innovations when introduced in markets. The cost reduction of supplied technologies is explained by the generation of know-how in technology development and standardization in manufacturing due to a larger scale of sales, called 'learning by doing'. The improvements made during the use of equipment prompted by more experienced organization are called 'learning by using'. The latter is particularly important for bioresources because technologies need to be tuned to specific ecosystems. The learning processes are cumulative, which means that the costs of technologies decrease in time as shown in Box 2.7. For instance, the prices of many natural resources decrease in real terms as a consequence of cost-effective exploitation technologies (Ahrens and Sharma 1997; Shafiee and Topal 2009, 2010).

Box 2.7 Technological learning is formalized as costs decrease in a period:

$$c_t = c_0\left(\frac{o_t}{o_0}\right)^a$$

For doubling of the production or sales a = 1/ log 2

c_t denotes the unit cost in year t and c_0 the initial unit cost; o_t the production of sales in year t and o_0 in the base year, and a is the learning rates which characterizes the inclination of the curve.

The decrease is measured by half-time; it is years until the costs are reduced by half. When the costs of photovoltaics decrease by 42%, it means a decrease to half of the cost per kW every 2 years; and when the costs of bioenergy decrease by 24%, it means a decrease to half every 5 years.

Learning rates of energy technologies (Rubin et al. 2015)

Energy technologies	Unit costs decrease per year	
	Minimum (%)	Maximum (%)
Solar PV	10	47
Wind Turbine	−11	32
Bioenergy	**0**	**24**
Pulverized Coal	5.6	12
Gas Turbine	10	22
Nature Gas Combined Cycle (NGCC)	−11	34
Hydropower (large scale)	1.4	1.4
Nuclear Power	Negative	6

2.4.3 *Sustainable Development*

The idea of economic growth in balance with social fairness and natural resources emerged under the term 'sustainable development'. The idea is that income can grow infinitely when innovations are steered toward fair distribution of wealth and environmental protection (WCED 1987). The steering capabilities, however, are debated. The concept of weak sustainability pursues solely corrections of income for damages to natural resources, for instance, the internalization of damages in the prices, whereas strong sustainability refers to the comprehension of welfare in a broader perspective than income; the latter is adopted by the United Nations.[4]

There is progress in the sense of lower impacts on environmental qualities and nature measured by materials consumption, energy and water use, forestry, pollutants, and waste (OECD 2003). An assessment for the 34 high income countries (OECD) during 1990–2010 shows relative decoupling in most countries, and an absolute decoupling in a few wherein the relation between income growth and more environmental impacts is sometimes negative and sometimes positive. A positive link could be due to the demands for better environmental qualities. Demands grow because ethical consumption, high-quality ecosystems, and the cultural expression of nature are increasingly demanded in addition to better use of natural resources and pollution reduction in production. Globally, the demands generate about USD 2900 billion for innovations that pursue sustainable development; it is 4.6% of the global GDP. A major barrier is policy support to vested interests, which is even larger than those demands (Krozer 2015).

2.5 Conclusions

Economics is a normative knowledge aiming at welfare in the sense of monetary and non-monetary values. Economy is comprehended as a system of circulating goods, created through the combinations of natural materials, labor, capital, and knowledge resources in exchange for money. Economies become less material-intensive and more knowledge-intensive due to a shift to services; further, thanks to the use of cleaner technologies and ethical products, lower environmental impacts per income, and sometimes in total, are observed when incomes grow. Welfare grows. Although bioresources have a small share in income in terms of money, they are a basis for valuable production and consumption, and provide a sink for pollution. These values are not expressed in money but can be monetized when the exploitation and degradation of natural resources are subjected to liabilities. The bioresources, herewith, possess advantages when compared to mineral resources in that they replenish spontaneously providing nutrients for new bioresources. It means that they are renewable

[4]http://www.un.org/sustainabledevelopment/.

resources that store and degrade noxious compounds rather than pollute. There is reason for optimism about sustainable development when liabilities are imposed and policies support sustainable innovations.

References

Ahrens WA, Sharma VR (1997) Trends in natural resource commodity prices: deterministic or stochastic? J Environ Econ Manag 33:59–74

Copeland BR, Taylor MS (2003) Trade and the environment: theory and evidence, 1st edn. Princeton University Press, Princeton

Daly HE, Cobb JB (1989) For the, 1st edn. Beacon Press, Boston

Dittrich M, Giljum S, Lutter S, Polzin Ch (2012) Green economies around the world? Implications of resource use for development and the environment. SERI, Vienna, Mimeo

Elkington J (1997) Cannibals with forks: the triple bottom line of 21st century business. Capstone, Oxford

Fischer-Kowalski M, Swilling M (2011) Decoupling natural resources use and environmental impacts from economic growth. United Nations Environmental Program, Mimeo

Grubler A (2012) Energy transition research, insight and cautionary tales. Energy Policy 50:8–16

ILO (2017) Global Wage Report 2016/17. International Labor Organization, Geneva

Krozer Y (2015) Theories and practices on innovating for sustainable development. Springer, Dordrecht-Heidelberg

Milanovic B (2012) Global income inequality by the numbers: in history and now. The World Bank, report nr. 6259, New York

OECD (2003) OECD environmental indicators, development, measurement and use. Paris

Piketty T (2014) Capital in the twenty-first century, 1st edn. The Belknap Press the Harvard University Press, Cambridge

Raworth K (2017) Donut economy. Chelsea Green Publishing, White River Junction

Rubin ES, Azevedo IML, Jaramillo P, Yeh S (2015) A review of learning rates for electricity supply technologies. Energy Policy 86:198–218

Sachs JD, Warner AM (2001) Natural resources and economic development, the curse of natural resources. Eur Econ Rev 45:827–838

Schumpeter JA ([1939] 1989) Business cycles, 4th edn. Porcupine Press, Philadelphia

Sen A (2009) The idea of justice. Allen Lane, Penguin Books, London

Shafiee S, Topal E (2009) When will fossil fuel reserves be diminished. Energy Policy 37:181–189

Shafiee S, Topal E (2010) A long term view on fossil fuel prices. Appl Energy 87:988–1000

WCED, World Commission on Environment and Development (1987) Our common future, 1st edn. Oxford University Press, Oxford

Chapter 3
Decision Making

Yoram Krozer

Decision making in economics is sometimes presented as if a law of nature, but it is about morals. Morals differ across societies and in time. Hence, only a toolbox for decision making can be delivered with regard to various, even opposing arguments and interests. The toolbox refers to private and collective interests because they differ, are often antagonistic in practice, even if aligned in theory. Decision makers can deliberate and negotiate using various tools, but it remains uncertain whether their intentions can be realized. In the process of decision making, resources and values are introduced, followed by assessment methods in production and consumption, transactions between customers and suppliers, and the basics of investments and valuation of nature. All those and exemplary accounts are shown with regard to bioresources.

3.1 Resources and Values

Economic decision making is about the possibilities of combining resources into the most valuable products with the lowest efforts; the efforts are usually measured as a cost and the value as a price. Resources are inputs transformed into things, services, images, models, and other outputs, here, called products. All those outputs that are useful and valuable for consumption are considered products, some eventually recycled; those that are useless are considered a waste. In economics, the demands for products are assumed to be unlimited; nonetheless, scarce resources pose limitations to the transformations.

Y. Krozer (✉)
University of Twente - CSTM, Enschede and Sustainable Innovations Academy, Amsterdam, The Netherlands
e-mail: krozer@xs4all.nl

Y. Krozer and M. Narodoslawsky (eds.), *Economics of Bioresources*,
https://doi.org/10.1007/978-3-030-14618-4_3

3.1.1 Resources

Products are created through combining natural resources, labor, capita, and knowledge. Usually several resources of each category are used, referred to as 'inputs', for realization of a product, referred to as 'output'. In theory, the number of outputs of any process is a factorial function of the number of inputs, which implies that several resources generate many possible outputs but only one is the most valuable product as illustrated in Box 3.1.

Several outputs are less useful, which means they are low value side products, need rework or recycling, while the useless outputs are considered as waste. The processes that deliver valuable products are referred to as 'effective', which means precise to the target. Costly know-how is needed to combine the inputs into an effective process. Since the costs of various resources differ, the lowest cost combination of resources is pursued, referred to as the 'efficient'. Given that effective and efficient processes can differ, the highest product value at the lowest cost is targeted, referred to as a 'cost-effective process'. A cost-effective process needs a lot of experimentation based on costly know-how; in some bioresource-based activities that can amount to as much as half of all costs, e.g., in pharma, personal care, designs, science.

Box 3.1 Outputs factorial of inputs

$N_o = N_i! + 1$

when N_o are outputs and N_i are inputs.

If three resources are involved, 7 combinations are possible: 3 * 2 * 1 + 1, but one is targeted. For example, an unspecified request for coffee can be served by only coffee, or milk, or sugar, or coffee with milk, or coffee with sugar, or coffee with milk and sugar, or only water. Milk, oat, sugar, and butter generate a targeted porridge with 25 possible combinations: 4 * 3 * 2 * 1 + 1. Any cultivation, for example, needs work (labor), seeds, shovel, rakes, can, knife, sack (capital), as well as soil, rain, and sun (materials). Without any knowledge, 10 resources generate 3.6 million combinations, but only 1 sequence of activities for harvesting crops.

Decision makers aim to maximize products, while reducing side products and preventing waste. Lower value side products can be recycled if resources are added, which implies an additional cost to the decision maker. Also waste disposal can be costly to the decision maker if the dissipation of materials is prevented. The dissipated materials as pollution cause damages, though the incubation time between disposal and damages can be long. Damages cause social costs. These costs are not covered by the decision maker unless liabilities for the damages are imposed on the individual sources of waste. When such liabilities are imposed, these individuals must pay the social costs of waste that are internalized in the prices of disposal. This way, social costs are converted into the private costs of waste disposers. The internalization puts the burden or responsibility of wasteful practices on the shoulders of the waste disposers rather than being carried by society. It makes wasteful transformations costly, rendering waste prevention an economic, it means cost-effective process. When liabilities on waste disposal are imposed the bioresources become advantageous compared to mineral resources because they are degradable and low-damaging. Their social costs are low.

The ideal of keeping products in economic circulation is launched, called the 'circular economy'. It envisages long-life products made possible by eco-design and recycling, but there are impediments. While eco-design can prolong the lifetime of a product, additional durability qualities can be costly. These products with a long-life are cheaper to use if they need low maintenance, but their longer life span impedes novelties. The main benefit of long-life products is fewer disposals; this needs the internalization of social costs. An alternative is recycling rather than disposal because the recycled products can be sold on the 2nd hand market instead of disposed of at a cost. This is economic in the case that sales minus their recycling costs exceed the costs of disposal. Again, the internalization causes higher disposal costs, which triggers recycling. Hence, the circular economy is driven by liabilities for damages caused by disposal and the possibilities to internalize these damages in the costs of disposal. The bioresources, herewith, usually have a shorter lifetime than mineral-based products because bioresources degrade faster but cause low damage. The internalization of social costs is a key driver of bioresources.

3.1.2 Values

The mainstream economics considers values as consumer preferences attached to products, called 'utilities'. When consumers are free to maximize utilities, they are sovereign; and when producers compete for the highest profit, the negotiations with producers generate the lowest possible prices. Such selfish consumers and producers generate efficient allocation of resources, meaning that any price change does not make anybody better-off. This is a stronghold in the mainstream viewpoint called 'perfect allocation' due to the 'invisible hand' of the free market. In this view, the free market generates social optimum, being maximum welfare. Individual preferences are assumed to be assessed based on opinions about willingness to pay for a product or accepted deficiencies. Such stated preferences would indicate the individual consumer demand for products, whereas social demands would be a sum of those individual demands corrected for harm to others.

The moral of selfish negotiators accomplishing social optimum is widely criticized (Sedláček 2011). The assumption about consumer sovereignty is even dismissed by the founding father of economic theory, Adam Smith, in 1776 (1986, p. 719): '*Consumption is the sole end and purpose of all production; and the interest of the producer ought to be attended to only so far as it may be necessary for promoting that of the consumer. The maxim is so perfectly self-evident that it would be absurd to attempt to prove it. But in the mercantile system, the interest of the consumer is almost constantly sacrificed to that of the producer; and it seems to consider production, and not consumption, as the ultimate end and object of all industry and commerce*'. Utilities are also disputed because consumers emulate the rich leisure class whose preferences are ceremonial rather than rational, expressed as consumer lifestyles (Veblen [1892] 2014). Furthermore it is observed that companies suggest free choice but supply deficient products; brands are celebrated as lifestyles though they are merely tools of sales; and, policies individualize consumers but issue obstructions to organizations that pursue the consumers' interests.

With regard to the utilities, the parity between private and social demands is also objected to. An observation is that individuals often pursue a private gain at the cost of groups' interests even if the pursuit of the groups' interest would provide better results to all, referred to as 'free riding'. It is also pinpointed that producers often seek entitlements for monopoly power from authorities rather than competition, referred to as 'rent-seeking'. A trade-off between individual and social interests is referred to as the 'prisoner's' dilemma'.[1] To elucidate the choice for individual interests: if two suspects of crime are separated for interrogation, there is a 75% chance that one of them will confesses or blame the other and only a 25% chance that they both deny having committed the crime they are suspected for.

The social dilemma is a variant that addresses decision making about a common interest. When two competing interest parties negotiate, usually referred to as stakeholders, they can decide for their private gain or for a collective gain. All factors being equal, the chances are 75% that each stakeholder chooses for its private gain and only 25% chance for co-operation between the stakeholders despite the gains of co-operation. It is 25% chance that no stakeholder wants to co-operate, 2 × 25% that only one of the stakeholders wants co-operate and 25% that both want to co-operate. Such a social dilemma is widespread in bioresources; for example, the choice between private gardens or collective parks, land for private housing or a natural area, connecting farmland or urban areas, as shown in Box 3.2.

[1]Look https://www.youtube.com/watch?v=uAcVrcrqhn8 and https://www.youtube.com/watch?v=TDtwsShOTjw.

Box 3.2 A social dilemma

If farmers want a road around the town for connecting fields and industries and to enable shorter transport to factories; given equal power and motivation, the chances are: 25% both decide not to invest, 25% industries do not invest, 25% farmers do not invest, only 25% both invest in a road that is beneficial to all, as shown in the table.

Chances of co-operation between two independent interests

Total chance is 100%	Farmers	Industries
Farmers	Yes/Yes = 25%	Yes/No = 25%
Industries	No/Yes = 25%	No/No = 25%

Co-operation is often beneficial. For instance, when one economy grows, all trading parties can benefit from that; when the debtor can repay a loan the lender and borrower flourish; when migrants add value and transfer money homeward both societies benefit, and so on. However, private gains and sentiments collide with the co-operative behavior. Co-operative strategies prevail when the co-operative behavior is awarded, while free riding and rent-seeking are punished by institutional arrangements, social conventions, or moral prescriptions (Ostrom 1990). Even altruism is generated when decision makers are awarded for socially satisfactory results albeit in an experimental setting (Simon 2006). Co-operative strategies usually generate higher benefits than those that are selfish.

3.2 Production

Production is realized, thanks to buildings, machines, stocks, and cash called ‘assets’; because payments are made and loans taken, called ‘liabilities’, and a payment to owners of the assets is attained, called ‘capital’. The assets are always equal to the liabilities plus capital on a balance sheet, called ‘debit’ and ‘credit’, respectively. Assets are generated when income from sales has been achieved, also known as proceeds, revenue, turnover, and so on. Liabilities must be considered when the cost of materials, machines, labor, wages, and rewards for capital are involved. Capital is generated when profit is made, or a net benefit after subtracting all costs is attained. The profits refer to commercial activities and net benefit to the non-commercial ones. The profits are divided among the shareholders, meaning owners of that specific production, the government and workers. The income of shareholders is called ‘interest’ or ‘rent’, the governmental income is tax and the workers income from profits is ‘*tantièmes*’. The net benefits are divided among stakeholders that are various interest groups.

When the total income equals the total cost, it is called 'break-even'. In production, break-even is expected to be the minimum performance because the profit or net benefit is nil. A profit implies that the income exceeds costs. The income can be generated during a period of time. The time that elapses before the total expenditure is covered by income is referred to as the 'payback'. The payback is the expenditure divided by the income necessary to cover it. If the income is counted per year, the payback is expressed in years and if the account is per month the payback is in months; for example, for an expenditure of 100 units and income of 20 units per year is a payback of 5 years, or 60 months.

3.2.1 Cost Function

A process is often expressed as a production function of output (product) with a set of inputs (resources). When the resources volumes are multiplied by the cost of resources, it is referred to as the 'cost function' as shown in Box 3.3.

Box 3.3 A production function is formalized as

$$O_t = f(N, L, S, K),$$

in cost units: N is natural materials, L is labor, S is capital, K is knowledge, O is output (of the product). Various letters are used because there is no uniform nomenclature. A cost function of 100 units furniture is, for example, 5 trees multiplied by the costs of a tree harvest (N), 2 carpenters times salary (L), 10 tools times purchases (S), and 1 expert multiplied by the cost of service (K).

Short-term changes in the production function are referred to as 'allocations'. An assumption in mainstream economics is that the output value increases when resource units are added. The incremental output becomes smaller as more units of that resource are added, assuming that all other resources are equal. This assumption referred to as the 'law of diminishing returns' is also called 'decreasing returns to scale'. A different observation refers to knowledge because an additional knowledge unit embedded in human skills and equipment can generate larger increment of output, it is increasing returns to scale. The increasing returns of know-how drive fast changes in production, for instance, the exponentially increasing agricultural outputs when pests are contained or exponentially decreasing costs of solar power technologies or energy storage.

Allocations of natural resources, including the bioresources, invoked many controversies in economic thinking and policy making. Mainstream economists assume that natural resources can be substituted for capital, for instance, land for greenhouses. A modernist argument states that natural resources as such do not exist because they are all human creations and given infinite knowledge, their exploitation can go on infinitely. Environmental economists, however, argue that natural resources are a non-reproducible foundation whose qualities need to be maintained for the future generations contrary to the reproducible labor, capital, and knowledge. Within the natural resources, minerals are non-renewable and bioresources are renewable under sound environmental conditions; maintaining of environmental qualities is a pre-requisite for bioresources.

It is also observed that cost-saving as a result of greater efficiency in production can be allocated in a larger production capacity. An efficient production can trigger more use of natural resources, referred to as the 'rebound effect' or 'Jevon Effect'. Hence, cheaper production can cause overexploitation of the natural resources; for example, cheaper machines can extract more wood from the forests. Nevertheless, allocations that involve an additional know-how generate more valuable products which counter potential rebound effects; for instance, exploitation of wood shifts from beam to plywood. Production of valuable products triggers the income growth and tempers the exploitation of natural resources, also of bioresources because less bioresources are involved, given income. Whether a bioresource is overexploited largely depends on the knowledge input.

3.2.2 Cost Categories

Some resources for production vary in proportion to the product's volume, also referred to as the 'scale of production'. The costs of such resources are variable, for example energy use. Other costs of resources hardly change with respect to the scale; these are 'fixed costs', for example buildings. Fixed costs increase at a rate that is slower than the pace of scale, while they remain similar when the scale decreases. There are also resources in use that were purchased in the past, whose costs are nil. These are referred to as 'sunk costs'. Sunk costs are often high because rails, roads, sewage, and other infrastructural work span many decades. The costs of bioresources are sunk when land or forests are possessed during decades. Many natural resources are not included in production accounts because they are assumed free of costs, for instance environmental qualities are considered free goods. However, they are invaluable and appreciated in time due to their unique qualities, this also hold true for biodiversity and the arts.

For example, in tomato cultivation, the costs of seeds and contract work are variable because they are divisible in proportion to the scale of production. However, some seeds are kept in stock for the next production period, called 'inventory', which constitutes a value that can be subtracted from the variable costs. The costs of shovels, rakes, cans, knives, and sacks are fixed because they remain even at nil production. These assets can be sold on second hand markets after discharge; the expected value of the sales, called 'salvage value' can be subtracted from the purchase of assets. The cost of constructing a road to the site of cultivation is usually a sunk cost because it was paid for in the past. The cost of knowledge about cultivation obtained from past studies is also a sunk cost, but when that knowledge is embedded in patents, it is an asset because it can be sold. Some environmental qualities are free goods but others are costs; for instance, sun for cultivation is a free good, but water and soil usually form variable and fixed costs, respectively.

The capital goods, ones that constitute the fixed costs, wear off with use or they lose value when kept idle. For example, shovels, rakes, and cans wear off slowly, in ten years; knives degrade faster, in five years, and sacks even quicker, say in three years. The loss in value of capital goods over a period is referred to as 'depreciation'. It is usually considered on a yearly basis. The value loss of patents, copyrights, and other knowledge-based goods is amortization. The remaining value of a capital good after depreciation is referred to as the 'book value'. It is the price of that capital good at that moment of time. The inventory and book value are quasi-price, also called a 'proxy', because they are the expected future prices of sales based on regulations about the lifetime.

The lifetime of products for depreciation is regulated by policies based on deliberations about performance of the vested products compared to novelties. An assumption is that novelties are cheaper and have a longer life but the lifetime of many products shortens because sales of novelties are pursued and consumers buy them. A shorter lifetime implies wasteful and polluting consumption. The residual value after years of depreciations, the salvage value, is the expected sales on 2nd hand markets, which is estimated right after purchase.

Capital goods are taxed. They depreciate based on regulations right after the purchases. Two depreciation methods are common. The linear method is popular and simple; it is the purchase cost minus the salvage value divided by the years of use as prescribed by the regulations. The accelerated depreciation is allowed in some countries for socially and environmentally benign technologies, such as renewable energy and ecological products. Fast depreciation facilitates investments due to lower taxes during the initial years. Both methods are illustrated in Box 3.4. The book value and salvage value are written off after the last year of depreciation. Thereafter, the book value is nil and the costs are sunk.

Box 3.4 Depreciation, total cost, and unit cost

Assume a machine that produces 1 million units, €1 per unit as variable cost, and the use of €1 million machine results in 5 years of depreciation, and €0.2 million salvage value.

Linear depreciation: It is $(1 - 0.2)/5$ is €0.16 million a year; the book value after 1st year is $0.8 - 0.16 =$ €0.64 million, after 2nd year $0.8 - (2 \times 0.16) =$ €0.48 million until the last year when it is written off together with the salvage value.

The total cost is €0.64 million + 1 million units × €1 per unit = €1.64 million, which is €0.164 unit cost or marginal cost.

Accelerated depreciation: The depreciation years is 100%, and the annual percent of depreciation is multiplied by accelerator, usually 2, which is (100/5)

* 2 = 40% annual depreciation. The purchase minus salvage value × 40% is depreciation after 1st year, i.e., 0.8 × 0.4 = €0.32; and the book value is 0.8 − 0.32 = €0.48. After 2nd year, the depreciation is 0.48 × 0.4 = €0.192; and the book value is 0.48 − 0.192 = €0.288... and so on.

The total cost in the 1st year is €0.48 million + 1 million units × €1 per unit = €1.48 million, it is €0.148 unit cost or marginal cost.

The total cost in the 2nd year €0.288 + 1 million units × €1 per unit = €1.288, it is €0.1288 per unit.

The sum of variable costs and depreciation is the total cost. The total cost divided by the production scale is referred to as the 'unit cost' or 'marginal cost'. Producers usually pursue the lowest marginal costs because it enables maximum profit for any given market demand. The lowest unit cost of production is considered cost-effective.

A profitable production is also referred to as the 'value added'. The value added equals the total sales minus the total costs of external suppliers, which usually refers to the costs of supplied materials. A high value added indicates that the natural resources are effectively used. When products are delivered in subsequent steps of activity starting from exploitation of natural resources, then the production of raw materials, followed by processing and distribution toward the consumption of products and finally disposal after its use—possibly after recycling—it is referred to as the 'value chain'. Each activity in a value chain adds to the costs of deliveries entailing the highest costs when products are consumed; if consumers are liable for the disposal costs, the highest costs are after disposal. The costs plus profits and liabilities reflect the market prices in value chains. Herewith, patents, copyrights, inventories, and other entitlements that support monopoly ownerships cause the market prices to surpass the costs and liabilities in value chains for instance, the prices of art, fashion, designs, medication, and other highly appreciated products are often many times higher than the costs in their value chains.

The costs of damages caused by pollution and degradation of ecosystems, which are the social costs of production, are rarely included in the costs of value chains. Markets externalize the costs of damages because no one is interested in accruing a liability for damages, unless these liabilities are enforced by authority or they can be traded as assets. The social costs of damages are factored into the costs of value chains when authorities enforce liabilities on those who cause the damage. For instance, emissions are traded in several countries because liabilities for those emissions are enforced. The costs in value chains including the profits and costs of liabilities for damages are referred to as the 'life cycle costs' or 'total cost of ownership'. When the liabilities for all expected damages are well-enforced, then life cycle costs reflect the total, private, and social costs of products.

3.2.3 Allocations

Whether to buy a new technology or keep the old one, for given demands, is indicated by the lowest total costs. A guideline in economics is to replace the vested technologies as fast as possible whenever new ones have lower total costs, even when the vested ones are hardly depreciated. However, note that the lifetime of depreciation is arbitrary and shortens because more products are introduced in markets and consumers discharge them faster as waste. Sunk costs indicate an obsolete equipment to be replaced because it involves costly maintenance or is less effective than novelties. For example, biogas in obsolete pipelines becomes costly when it leaks or the pipelines must be maintained, and the sunk costs of obsolete pipelines obstruct cost-effective alternatives, such as biogas for the local co-generation of heat and power. Similarly, the sunk costs of obsolete sewage impede innovative wastewater treatment, such as the distributed sanitation with digestion. The simplified deliberation of the replacement is shown in Box 3.5. The total costs of vested technology are the variable cost per unit multiplied by the production scale plus the book value minus the salvage value. The total costs of new technology are the variable costs per unit multiplied by its scale plus its purchase cost. The accelerated depreciation enhances that replacement.

Box 3.5 Technology replacement

Technology replacement	Old machine	New machine
Production cost of one unit	10	5
Number of units	1000	1500
Production years	5	5
Total variable costs	50,000	37,500
Book value	10,000	20,000
Salvage cost	−1000	
Total cost	59,000	57,500
New–old		−1500

Conclusion: Replacement is Cheaper

Another issue for decision making is whether to specialize or diversify. The specialization refers to allocation in a particular activity rather than spread across various activities. A guideline in an economic system is productivity of a sector or business, and in decision making it is profitability; the profitability means the highest profit per key resource. The highest profitability acts as a guiding priority for allocation. The selection of a key resource is somewhat arbitrary because it depends on production characteristics. For instance, a key resource in the production of bioresources on farms is usually land area, but in industries it can be material or labor. Box 3.6 illustrates an allocation between two departments in a firm.

Box 3.6 Allocation for the specialization

Given limited space of a wood center, what service should be expanded? A wood center on 7000 m^2 at €100 per m^2 has the storage service on 5000 m^2 with 2 staff which generates €0.7 million income and the information service on 2000 m^2 with 1 staff which generates €0.3 million income. Each staff costs €50,000 a year.

Productivity, all × 100	Storage	Information
Space use in m^2, €100/m^2	5	2
Space cost in €	500	200
Staff	100	50
Total	600	250
Income	700	300
Profit	100	50
Profit €/space in m^2	140	**150**

The information service is more cost-effective and should be expanded

Costs also reflect the scale of production. The total costs increase with scale, but the unit cost usually decreases. If all costs of resources are perfectly divisible with respect to scale, as observed in many services and assumed in mainstream economics, then unit costs decrease in a linear manner along with production scale called 'homogenous cost functions'. However, when resources are inflexible because they involve technologies embedded in equipment and constructions, the costs are imperfectly divisible and the unit cost decreases slower as production scales up, called 'heterogeneous cost functions'. The latter is typical of agriculture and industry wherein fixed costs are large. Herewith, unit costs are estimated based on scaling factors which relate the scale to units costs as shown in Box 3.7. The scaling factors are determined based on experiences.

Box 3.7 Scaling

The scaling factors are estimated as follows

$\alpha = ln(c_j/c_i)/(P_j/P_i)$

When α is a scale factor, $c_{i,j}$ and $P_{i,j}$ are marginal costs and the production scale, respectively. For example, if the scale increases 2 times and marginal (unit) costs 1.5 times, the scaling unit is ln (1.5/2) = ln (0.75) = −0.29.

The smaller the scale factor, less divisible is the technology.

3.2.4 Production of Bioresources

All principles of allocation also hold true for bioresources. In addition, variation in the climatic and ecological conditions influences decision making about cost-effective allocations. Plants grow spontaneously with sun, water, and nutrients, but the environment poses limits to that growth.

If nutrients in the soil are available for bioresource production, sunlight and water are limiting factors. Although the total solar energy that reaches the surface of the Earth is thousands of times greater than our present energy consumption, its intensity per area varies. The spread is from 100 kW/m^2 in the Arctics with much water; 1500–4000 kW/m^2 with sufficient water in temperate and warm climates, respectively, and up to 6000 kW/m^2 in deserts with little water. Plants grow mainly in temperate and warm climatic conditions due to the balance between the sunlight and water.

Plants need soil nutrients for their growth, referred to as ‘primary production’. They also form nutrients for subsequent generations of plants as well as for animals referred to as ‘secondary production’. Organisms reproduce within an ecosystem, given the limits posed on them by their environment. A large production of bioresources in an ecosystem, one that exceeds the primary and secondary production, needs additional nutrients from resources external to that ecosystem—for instance, mineral fertilizers. Vice versa, when too many nutrients are added and not absorbed within the ecosystem, they dissipate as pollution. Allocations in the production of bioresources must strike a balance between scaling up, inputs of external resources and regeneration of soil nutrients within an ecosystem.

Globally, the scale of bioresources in ton cereals per hectare enlarges due to increased use of fertilizers, pesticides, and other materials and machines. However, the unit costs increase even faster because of costly expenditures. The product value in terms of price per ton cereals also increases, but at a pace that is slower than that of scale which leads to lower productivity being measured by the monetary value per resources mass. A larger scale does not help increase productivity, rather the downscaling of technologies does. Larger expenditures and mergers of companies are pursued for their private advantage but they rarely contribute to cost-effective production or national economies. The productivity growth in the sector saturates and the profitability of individual companies drops.

Alternative strategies are also pursued: downscaling of technologies. Efficient use of energy, nutrients, and other resources is fostered, known as 'low-input agriculture'; nonetheless, there are limits to the cost-saving even if production moves to low cost countries where energy and chemicals can be partially substituted for manual works. High value specialities increase productivity but require costly research and development into efficient production methods and novel varieties of plants and breeds. Diversification of services into agriculture, forestry, and tourism adds value but various types of know-how are needed. Dedicated 'fair', 'eco', 'healthy', and suchlike attributes in products serve the consumers demands for ethical and esthetic purchases, but the scale of sales is small. All alternatives have pros and cons, whereas mixes are laborious because they involve opposing activities, for example, cost-saving of resources often contradicts diversification.

A decision maker can compare alternatives based on opportunity costs; note that a negative cost is a benefit. For example, the production of bioresources delivers products for sales and organic residues which can be used for soil enrichment, fodder, in industries, as biofuels, and for other purposes. The use of residues for soil enrichment that avoids use of mineral fertilizers can be compared to the costs of mineral fertilizers, or to the use and costs of biofuels. The cost of soil enrichment with the bioresidues can be lower than the costs of minerals but higher than the costs of biofuel based on those bioresidues because can replace diesel. In all cases with regard to decision making, the lowest opportunity cost should prevail which has been exemplified in Box 3.8.

Box 3.8 Opportunity cost

Alternatives	Mass	Unit cost	Total cost	Integral	
Cost					
Residue reuse	10	0	0		
Mineral use	10	+5	50	*+50*	Opportunity cost
Income					
Biofuel use	10	−10	−100	*−100*	Opportunity cost

The lowest opportunity cost is the biofuel

The opportunity costs of the present, intensive production of bioresources are high. The social costs of displaced labor, degraded soil, polluted water, and endangered biodiversity can be mitigated when a smaller scale of production generates a higher value from ethical and esthetic attributes, agro-tourism, leisure, and other services.

The alternatives also reduce the risks of liabilities for social costs and dependency on subsidies. Such alternatives in the bioresources production emerge and are labeled as ‘organic agriculture’ or the like. For instance in the European Union, the area of intensive agriculture decreased, whereas the area of organic agriculture increased from close to nil in the 1980s to 7% of the total area in 2016 which is—21% in Austria, 18% in Estonia, 14% in Czech Republic, and Italy but only 2% in Ireland and Romania. The share of organic agriculture grows by an annual average of 4% in Europe, when compared to hardly any growth of intensive agriculture; the food produced by ‘organic agriculture’ and labeled ‘organic’ covers about 6% of all foods, translating to about USD 60 billion a year. Although its output per hectare is substantially lower the profit per hectare is often similar because the output value is higher and costs of mineral are lower.

3.3 Consumption

Free market is often put up on a pedestal as an ideal. It is assumed that the transactions between consumers aiming to satisfy their individual preferences and competing product suppliers aiming at maximum profit determine the lowest prices. Since higher prices are assumed to reduce demand, and vice versa, the prices indicate the consumers’ preferences and such transactions between individuals are considered superior to non-market transactions between institutions. Such collective transactions are labeled as ‘market interventions’ which are perceived as necessary for the safety, ownership, and other general interests of the market but at the same time are considered the cause for market imperfection.

3.3.1 Prices and Qualities

In the ideal of market transaction, selfish demanders and suppliers maximize their utilities and profits, respectively. Due to such selfish behavior, they reach equilibrium in prices when suppliers’ marginal costs equal the consumers’ marginal benefits. Such prices generate the highest welfare in the mainstream economic view. In practice, however, millions of consumers purchase products in supermarkets, traders exchange in millions of tons, banks transfer billions of dollars, and all with the presupposed notion of trust that these transactions are fair. Integrity in the transactions implies that the stakeholders are not ripped off or robbed even if an individual stakeholder can gain from it.

Markets operate by trust fostered by cultural conventions and enforced by, authorities, and set by institutional agreements. When the market prices deviate from the theoretical equilibrium prices because producers or consumers pursue their self-interest, they are countered by the institutions that watch and guard for the sake of trust; or the law is enforced when the deviations are excessive as perceived by those institutions. The perceived deviations are referred to as 'surpluses'. The producers' surplus is taken as their sales price minus the lowest supply cost; while the consumers' surplus is their willingness to pay for the product minus the sales prices; all prices are multiplied by scale as shown in Box 3.9. Based on this assumption energy producers, telecom companies, media, and other businesses are accused of unfair prices because of monopoly positions on markets. Unfair transactions or untrustful institutions undermine economies; integrity is key for prosperity.

Box 3.9 Example of the producers' and consumers' surplus

Imagine that the price is USD 5 per unit. Producers deliver 100 units at USD 3 per unit. The producers' surplus is USD $(5 - 3) \times 100 = 200$. The consumers are willing to pay USD 8 per unit. The consumers surplus is $(8 - 5) \times 100 =$ USD 300. Total surplus is USD 500.

Imperfections are inherent to market negotiations referred to as 'asymmetric information'. It is because each subsequent supplier in a value chain compounds qualities into its product at an additional cost and profit when aiming to satisfy its customer demand. Further, every customer bears a higher price if higher performance is perceived though it can be found deficient after purchase. Herewith, suppliers determine the product qualities and are aware of deficiencies unknown to customers, because the latter possess an interest to demonstrate their preference, whereas suppliers have no interest to highlight any product deficiency. In addition, trade-offs between product qualities are observed. The perceived performance refers to functional qualities, which means usefulness of products, and ethical values comprise expressions of social conventions. For example, the former can be strong and durable and the latter can be natural and fancy. The suppliers' costs and profits add up in value chains but the supplied qualities do not because of rivalries between the functional qualities and ethical values, as well as within each of them. For instance, stronger products can be heavy or ugly, or fancy products can be fragile. Hence, products can be of poor quality—even damaging—without information about such trade-offs, discovered only post purchase. The suppliers' liabilities for the deficient performance would prevent imperfection but cause high transaction costs and they are resisted by the suppliers.

The observed imperfection in market responses to prices are expressed as 'elasticities'. Elasticity relates the purchased volume to prices as shown in Box 3.10. Demand elasticity relates an increase in the price to a decrease in purchases: products are called price elastic when an increase in their price causes a larger decrease in the product volume. Then, the price elasticity is below -1. Products are called price inelastic when an increase in their price causes a smaller volume decrease. In this case, the price inelasticity would be between 0 and -1. Luxury products are often price elastic, basic products price inelastic.

Box 3.10 Price elasticity of demand

The midpoint of elasticity is:

$$E = \frac{p_{i+1} - p_i}{(p_{i+1} + p_i)/2} \Big/ \frac{q_{i+1} - q_i}{(q_{i+1} + q_i)/2}$$

E is elasticity, $\frac{p_{i+1} - p_i}{(p_{i+1} + p_i)/2}$ is percent change of price, $\frac{q_{i+1} - q_i}{(q_{i+1} + q_i)/2}$ is percent change of quantity, p_{i+1} is subsequent price of p_i and q_{i+1} are the subsequent quantity of q_i.

For example for $p_i = 4$ and $p_{i+1} = 12$, whereas $q_i = 100$ and q_{i+1} 60 the following can be calculated:

Price change $\frac{12-4}{\frac{12+4}{2}} = \frac{6}{8} = 0.75$; quantity change $\frac{60-100}{\frac{60+100}{2}} = \frac{-40}{80} = -0.5$; elasticity is $0.75/{-0.5} = -1.5$.

Price inelastic demands can be caused by perceived deficiencies, opinions biased by conventions, limited capabilities, and other impediments in response to prices which cause habitual transactions rather than deliberated ones. Some valuable products have positive price elasticity. For example, the purchase of energy resources is negative inelastic, but household purchases of electricity enlarge despite their higher prices than fuels in many countries. The price elasticity of demands also depends on income. The purchases are usually a U-like function of income: low income and high income are generally more elastic than mid-income, because the former needs more products and the latter can afford them. The price elasticity of supply links an increase in prices to supplies. It varies from negative to positive; the latter is observed for its ethical attributes, for instance, fancy cars. Cross-elasticity indicates a substitution between products, in response to given prices; elastic products indicate a growing market share when prices decrease.

3.3.2 Non-market Transactions

Many qualities are not expressed in prices; for instance nature, nurture, joy, and quarrels have no prices. In general, prices are absent or they are deficient when transactions address feelings (e.g., security, friendship), conventions (e.g., esthetics, wisdom), collective interests (e.g., climate, species), and when large time-lags are involved (e.g., arts, science). Qualities that are not expressed in prices but serve society or ecosystems as a whole are referred to as the 'common goods' or 'public goods'. It is a common good when the uses of that good are rival, for instance species, or it is public goods when the uses are non-rival, because used by many side by side such as parks.

The unintended side-effects of transactions on other interests, external effects, are also referred to as 'impacts'. The positive external effects are often accommodated in market transactions but are rarely paid for on markets; for instance, knowledge is used for new businesses but rarely paid for by start-ups. Negative external effects are rarely accommodated spontaneously but enforced by regulations because of a lack of interest, for instance pollution. Pricing of common goods can even aggravate negative external effects. For instance, it is widely assumed that high prices of bioresources will reduce the demand for scarce bioresources, but wildlife is poached and ecosystems trampled upon when highly priced because they have come to be known as exceptional and liabilities for damages are evaded or not enforced.

The prices of common and public goods as well as external effects are framed by political processes. However, these processes cause biased decision making when private companies are entitled to act as an authority and public organizations operate as private firms because it is attractive to earn as a private firm and spend as an authority. Such blurred lines corrupt decision making entailing costly and deficient implementation. For instance, entitlements for large-scale cultivation of bioresources for fuels trigger excessive costs, as well as degradation of nature and pollution. When free riding is allowed, moral integrity in decision making is undermined.

Alternatives are pursued. Public institutions regulate the use of common goods and enforce liabilities for their degradation, for instance, foster reforestation and, respectively, penalize deforestation. Conventions about common goods between private organizations are sealed, for example, about the bioresource production versus protection of nature in national parks. Transactions also become more transparent when local value chains are pursued and dedicated suppliers rather than auctions are prioritized. Consumers are socially responsible when they demand ethical attributes in products (Devinney et al. 2010), so also when producers eco-design their products. All these alternatives foster development and the use of bioresources.

Valuable uses of ecosystems are referred to as 'ecosystem services'. These services are usually divided into four categories. The category 'supporting services' refers to services that support nutrient cycles and soil formation. Based on which plants and animals enable the regulation of climate, decomposition of waste, purification of water and air, pest control, and suchlike, referred to as 'regulating services'. Supporting ecosystem services also provide energy, foods, materials, pollination, genetic resources, and other products, called 'provisioning services'. Finally, ecosystems services that refer to spiritual, historical, recreational, and other cultural uses are called 'cultural services' (Millenium Ecosystem Assessment 2005). The value of ecosystem services is estimated based on willingness to pay similar to if they were market services. Table 3.1 summarizes the minimum willingness to pay for ecosystem services based on many studies, scaled up to the global level (Constanza et al. 1997).

Such a valuation of nature is disputable because it reflects conventions, rather than opinions; it is rarely revealed in behavior as a payment for a service (Kahnemann and Knetsch 1992). The assessments aim to evaluate all ecosystem services, but are biased when particular ecosystem services get little attention; for instance, the cultural services (Milcu et al. 2013). Their relation to prices is also dubious; for instance, should the terrestrial services be offered for agriculture because its USD 760/ha is lower compared to agricultural production of USD 2360/ha in Europe?

Table 3.1 The minimum willingness to pay for ecosystem services on a global scale; the main bioresources are shown in bold (Constanza et al. 1997)

Ecosystem services	Million ha	Value in USD/ha/year	USD Billion/year
Marine			
Open Ocean	33,200	252	8381
Coastal	3102	4052	12,568
Terrestrial			
Forest	4855	969	4706
Grassland/Rangeland	3898	252	906
Wetland	330	14,786	4879
Lakes/Rivers	200	8498	1700
Desert	1925		
Tundra	743		
Ice/Rock	1640		
Cropland	1400	92	128
Urban	332		
Total	51,625		33,268

Economists therefore advocate the use of physical indicators for ecosystem services in addition to market transactions based on prices, for instance, land use in market prices and biodiversity in species. This viewpoint is adopted in the assessments of ecosystem services.

3.4 Investment

Expenditures aiming to generate capital in the future are considered as productive; this is contrary to the consumptive expenditures for current uses. The productive expenditures are referred to as 'investments'. Investments are done by companies that buy machines, financiers that purchase companies' shares, as well as by consumers that refurbish houses for rent or make jam for sales. All those investors need finances, supplied by participation or financing. Participation is when an investor obtains equity from family, friends, crowd funding, venture capitalists, shareholders, and other participants. Subsidies from the state add to the equity even if the state does not participate. Equity gives rights for a share in profits called a 'dividend'. Financing is when an investor obtains loans from banks, funds or other services which must be repaid with interest. The financiers can also demand a guarantee for the repayment, such as house, farm or other capital collateral. Tax exemptions, guarantees and such like facilities are instruments of the state aiming to reduce the financing burden of investments.

3.4.1 Present Value

An investment implies risk taking because it is uncertain if that expenditure will be covered by incomes from production in the future. The investment decisions are usually based on experiences from the past about the costs versus the aspired to future income; and the investment decision cannot mitigate uncertainty about the future income. Hence, such decision making is not solely a rational deliberation about costs and income but driven by greed, desires, public opinions, and other factors that concern the attitude called 'bounded rationality'.

The income of investors and financiers is interest, also called 'rent' as a reward for delivery of capital and compensation for taking risks. Interest should cover all expected risks of expenditure. The risks are that the investment cannot be properly repaid because recipients of capital fail to perform, as well as inflation, devaluation, furthermore changes in policies and consumption patterns and other factors that cause uncertain income in the future. Since the value of money in the future is lower than its present equivalent, idle capital is considered a cost. That cost is expressed as a percentage charge per period, referred to as 'cost of capital' or 'interest rate'. It is, for instance, 5% interest rate a year on a bank, or 0.5% a month personal credit which means 0.5% × 12 months is 6% a year. The interest rate is compounded per period of time; the time is usually counted in years for investments, in months for personal credits and in days on a credit card.

The interest rate cumulates per period of time. Investors therefore must consider the interest rate and period of time. For a profitable investment, the investors' expenditures, also called 'outgoing cash flows', must be covered by income, the 'incoming cash flows', during a period of time. The value of present asset at a moment in the future is Future Value (FV), for example, repayments of a loan, given an interest rate, or value of a house when fees for rent annually increase. The future value function of time, given the interest rate, is shown in Box 3.11.

Box 3.11 Future value

$$\mathrm{FV}_t = \sum_{t=1}^{n} \mathrm{C} \cdot (1 + r/b)^{bt}$$

FV is future value, C is principal of the investment, r is the interest rate, b is number of payments per period, and t is periods.

When a financier lends 1 million euro (principal) at 10% interest and annual payments future value is:

1st year: 1 * (100% + 10%) = 1.1 million euro
2nd year: 1 * (100% + 10%) * (100% + 10%) = 1.21 million euro.

Since idle capital or delay of investments implies a potential loss in money, the values of assets decrease in time. The present value of incoming cash flows in the future decreases at the interest rate per period referred to as 'discounting'. The discounted incoming cash flows during a period of time at an interest rate can be summed up to the present value (PV). When the present value of all outgoing cash flows during a period of time, it is the flow of expenditures, is subtracted from the present value (PV), then net present value is obtained (NPV). The calculations of PV and NPV are shown in Box 3.12. NPV indicates profits that are generated due to the investment after the discounting. The net present value is a key indicator for the potential profits of a firm or benefits in society.

Box 3.12 Present value

$PV_t = \sum_{t=1}^{n} C/(1 + r/b)^{bt}$

When 1 million euro is the outgoing cash flow, investment, I, generates 1 million euro incoming cash flow, income, C, at 10% interest the net present value is.

1st year $1/(100\% + 10\%)^1 = 0.91$
2nd year $1/(100\% + 10\%)^2 = 0.83$ million euro
PV = 1.74
NPV = PV − I
NPV is 1.74 − 1 = +0.74.

Positive NPV indicates potentially profitable investment, negative NPV a loss.

When several investments in time are done, they should be discounted in the same way as the income.

Businesses also use Internal Rate of Return (IRR). IRR is the average interest rate to attain break-even, it means when income covers investments; the interest rate is found iteratively for nil profit (i.e. income minus investment). Although IRR can cause inconsistent results and is less precise than NPV, it is convenient because it provides one number of interest rate for the decision making.

When investors deliberate alternative proposals with different periods of time for the expenditures or different time frames of incomes, they can equalize the duration of the alternatives or estimate the present value of average annual cash flows. The latter is usually convenient. Such an average cash flow is called the 'uniform annual stream' (UAS), 'equivalent annual cost' (EAC), 'present value annuity', or names like such. It is calculated as the present value or net present value divided by the annuity factor. The annuity factor is the sum of discount factors during the project lifetime. See Box 3.13.

Box 3.13 Uniform annual streams

The annuity factor A_t is the sum of the discount factors over time, and the uniform annual streams (UAS) is NPV (or PV for nil investment) divided by A_t:

$A_t = \sum_t^n 1/(100\% + r)^t$
UAS $=$ NPV$/A_t$

Based on the above (3.12): $A_t = (100\% + 10\%)^1 + (100\% + 10\%)^2] = 1.1 + 2.21 = 2.31$.

UAS = 0.74/2.31 = 0.32 average profit a year.

The uniform annual streams are often used when investments in large-scale projects are compared, for instance in the development of housing, utilities, or infrastructural works. This is done because the durability of project alternatives differs, whereas the investments and incomes are large. Various formulas for the financial assessment can be found in Excel.

3.4.2 Cost–Benefit Assessment

Since the private costs of a company or consumer differ from the social costs of a community, region, country or another collective, it is desirable to consider all costs and benefits of individuals as well as ones of the collective; for instance, when various stakeholders need to agree on investment. A cost-benefit assessment aims to provide objective results for decision making about large investments taking into consideration all private and public interests. In theory, all expected costs and benefits directly related to the investments should be estimated, including the expected costs of damages. Then, the benefits, costs, and damages are discounted at an interest rate that reflects societal risks. These are typically lower than private risks because society can bear more risks than individuals and because the states and other collective organizations can obtain loans at a lower interest rate than the individuals. A positive net present value or uniform annual stream indicates an attractive investment as shown in Box 3.14.

Box 3.14 Cost-benefit assessment

$$\text{NPV} = \left[\sum_{t=1}^{n} (B_t - C_t + D_t) \cdot r^{-t}\right]$$

B is total project benefit; C is total project cost; D is total damage cost; r is interest rate; t is the lifespan of the assessment.

Many handbooks on cost-benefit assessments can be found (e.g., Boardman et al. 2010; Brent 2006; Campbell and Brown 2003; Zerbe and Bellas 2006) but the method is much criticized. First, data on the costs, benefits and damages are uncertain. Second, even if sound data is used, some stakeholders may perceive a benefit while others deem it a cost; for instance, taxpaying organizations may perceive different benefits than tax spending ones, or a region can gain a benefit but the country, a loss. Such distributional deficiencies cause biases in favor of a specific stakeholder. Third, the interest rate for discounting is disputed. Since the Present Value of costs and benefits in the long term evolve to nil, the short-term benefits are deemed attractive. The discounting therefore causes myopic decision making, meaning a bias for short-term gains. Damages and benefits in the long term are underestimated, for instance, in the protection of nature.

The methods of monetization of benefits and damages are also debated. The contingent valuation methods refer to the willingness to pay for quality, or acceptance of compensation for degradation and to co-joint methods which combine various alternatives. However, the contingent valuations express vested conventions rather than individual preferences and demands. Other methods reveal preferences based on statistical data. Hedonic pricing correlates the purchased qualities to payments, for instance, housing prices to distance from nature. The travel costs methods compare the cost of visits to alternative destinations. The damage costs methods monetize damages, such as additional costs of care, repairs, and others. The value of life methods assesses payments for risky activities or the costs of preventing causalities. Several methods aim to valuate ecosystems services. The avoided cost and replacement cost show the cost of maintaining an ecosystem and the costs of alternative projects, respectively; for example, biological versus chemical pest control. The opportunity cost method shows the costs of alternative locations, for instance, a park or housing. The production function shows the income from using ecosystems, and the net factor income shows the income after subtracting the costs. However, the results of such analyses are sensitive to data manipulations and gaps in causalities.

Experiences show that costs often grow beyond estimates in the cost-benefit assessments, damages are insufficiently acknowledged because of scientific findings or entry of new stakeholders and benefits are disappointing because better alternatives are not included in decision making or emerge thereafter. Even an excellent cost-benefit assessment, therefore, is not a scientific verdict about investment but a tool in support of deliberations. Cost-benefit assessment is not a panacea for decision making about bioresources. That said, normative deliberations and negotiations between stakeholders can be a more transparent and democratic process of decision making. Deliberations can be based on physical indicators for benefits and damages, eventually monetized if data are transparent and uncertainties are acknowledged.

3.4.3 Interest Rates

In mainstream economics, interest rate is considered a resultant of transactions between profit-maximizing demanders being anyone from consumers, businesses, to countries and financiers that supply capital. These financiers are banks, insurances, pension funds, as well as wealthy persons and their funds. The interest rate is high when capital is scarce because much is demanded. It increases when financiers perceive high risks because income is uncertain or the economy fragile. Regarding risks, the annual interest rate varies from 0% for savings guaranteed by the State, bonds which are loans to the State, securities which are loans to funds, loans to companies, loans to persons, shares in companies, participation in new ventures, and up to 50% and more for participation in know-how. In general, higher interest rates are demanded for equity than for loans, and innovators rarely get loans because it is considered risky.

That viewpoint about market transactions is disputed with regard to the accumulation of capital. Due to high interest rate, when it is well-above economic growth, capital accumulates in the hands of financiers. When the financiers merge, competition on the financial markets is undermined and substituted for rent seeking through entitlements, and speculations; these are policy support of failures and financing without assets, respectively. Capital accumulation undermines economies when that rent-seeking is favored by individuals and collectives above financing investments. If penalties on the financial markets are low because policies accept or foster rent-seeking, financial bubbles are generated entailing crashes and economic depressions—as observed throughout centuries (Ferguson 2009).

Interest rates are accepted in modern economies but contentious in religions and ideologies because of the accumulation process. Concerns about the interest rate are found in many cultures. For instance, Judaism prescribes a limited period of gains from loans followed by eradication of all debt within fifty years. Interest is banned by Christianity and Islam in favor of voluntary offerings to the Church and Mosque; Jews were allowed to operate with interest and were therefore persecuted. Also, in the socialist school of thought, there are objections made to interest because capital accumulation undermines the distribution of wealth. Capital accumulation is considered contradictory to the meritocratic principles of liberalism because it provides gains to the wealthy, rather than the gifted. In the environmentalist argumentation, interest rates drive overexploitation of natural resources because the present value of future resources is discounted.

Alternatives to an interest rate, however, can be even more disastrous as seen in robbery by the state and mass murder for lack of capital. For instance, the voluntary offerings that turned into obligations to the Catholic Church ignited the religious wars of Reformation and Post-Reformation in Europe, which decimated its population. The communist ruling banned rent but imposed mass domestic slavery in order to compensate for that deficit. Nazism even exterminated the Jewish population based on accusations of capital accumulation and other nonsense, and Facism imposed slavery on millions.

Restrictions on capital accumulations could do better. Policies can correct failures of the capital market by setting stringent standards for credit. It can also set a low interest rate for bonds called 'discount rate', which acts as a minimum cost of capital. Policies can increase supply of money entailing lower interest rates, albeit at the cost of inflation. Rather than purchasing malicious assets or liabilities of financiers as done by the Central Bank after the financial crisis in 2008 labeled as 'Quantitative Easing', governments can shift the risks of liabilities from the public to shareholders and expropriate those failing financial institutions. Taxing capital transactions—called 'Tobin tax'—even if small hinders rent-seeking. Personal charges for disinformation and gambling foster trust in financiers. An array of policy instruments can contain financial crashes, but they are insufficiently used or even dismissed in priority of rent-seeking.

In case of bioresources, a lower interest rate is sometimes accepted for the common good. It is if an asset is expected to be sold at a premium price, referred to as 'goodwill'; for example, investment in a tourism resort at nature reserve has goodwill because it is linked to a common good with high public value. Lower rates are advocated for investments in common goods. Assuming goodwill, some financiers are focused on investments with ethical attributes called 'impact investments'. These are investments in education, drinking water, renewable energy, biodiversity, and other common goods. Although the global volume of impact investments is small—USD 114 billion which is only 0.2% of all assets—it grows fast.[2] Policies reduce risks of the impact investments through community banks and encourage private financing with tax exemptions, soft loans, and guarantees. The risks are also reduced due to purchases, labor, knowledge support, and other instruments.

3.5 Value of Nature

Nature is a common good whose qualities degrade when elements are separated and privatized. Nevertheless, exclusive rights are pursued. For instance, space, climate, and atmosphere are claimed for satellites, geo-engineering, and flight strips, respectively. Seas are captured as territorial waters and continents divided by borders. Mountains, deserts, wildlife, and wasteland are claimed by community; forests and nature reserves are claimed by authorities, species are owned in gene banks, parks, zoos, and the like. If gains due to those rights are maximized for self-interest but liabilities remain unsettled, the free riding undermines nature entailing degradation of ecosystems and environmental qualities. The impact of free riding also holds true for bioresources when harvesting exceeds the reproductive capabilities of nature.

3.5.1 Sustainable Harvest

A sustainable harvest can be assessed based on the balance of growth and death in the population. Per period of time, the growth rate of organisms without constraint is generally between 0 and 1; zero growth rate means that birth and death are precisely in balance and 1 implies a doubling. Bacteria can double nearly every hour, which implies about 10 million individuals in a day if there are no limiting factors. Annual

[2]http://www.undp.org/content/sdfinance/en/home/solutions/impact-investment.html.

plants reproduce within a year and the perennials within a few years. While the global human population is large, yet the annual growth rate has averaged at only 1.2% throughout the last century, contrary to the idea of a population bomb. For given ecological conditions, the reproduction of a population is defined as the net increase of the birth of organisms over deaths in a defined period of time as shown in Box 3.15.

Box 3.15 Reproduction of populations

Number of organisms grows exponentially if environmental conditions do not impose limits:

$N_t = N_0 \cdot e^{rt}$ or $dN/dt = rN$, e Euler number

Within the limits, the net enlargement of a population per period is:

$r = (B_t - D_t)/N_t$

r is the growth rate; B is the number of organisms born, D is the number of dead organisms, and N is the number of organisms in a population at t the moment.

The population growth is limited by scarce food, space, predators, and other ecological conditions referred to as the 'carrying capacity'. Populations, measured by number of organisms, usually grow as a sigmoid function of time (S-curve): it grows fast initially, then saturates and finally declines because the carrying capacity is approached and ultimately surpassed. The growth rates of any sigmoid function are represented by a parabola; the rates increase to a maximum and decrease when the total population approaches the said environmental constraints. The highest point on the parabola is the maximum sustainable harvest. It is when the highest rate of harvest equals the highest rate of regeneration; more harvesting is unsustainable and less is not at the maximum harvest, which is formalized in Box 3.16.

Box 3.16 Assessment of the maximum harvest

Given maximum population K related to environmental conditions is:
$dN/dt = rN \cdot (K - N/K)$

The population growth rN is followed by saturation when number of individuals N approaches K.

Calculations using spreadsheets can be found on YouTube.[3]

[3]https://www.youtube.com/watch?v=c6pcRR5Uy6w, https://www.youtube.com/watch?v=rXlyYFXyfIM, https://www.youtube.com/watch?v=6bd3kUHdiUI, https://www.youtube.com/watch?v=uemhtqZHnak.

3.5.2 Economic Harvest

The optimal economic harvest differs from the maximum sustainable harvest because costs and income are considered in light of the profit. At any price, income follows the harvest line on the parabola but costs increase as the harvest grows. Figure 3.1 shows the costs if they increase in proportion to the harvest volume.

The income declines after achieving of the maximum sustainable harvest. The costs, however, grow when even more is harvested. This implies that the maximum profit can be obtained below that maximum, just left of the highest point on the parabola (Perce and Turner 1990). From the community perspective, it is profitable to maintain efforts below the maximum sustainable harvest unless the costs do not decrease with the harvest due to technological changes or a larger scale of harvesting ensues. In the interest of an individual, however, it is attractive to capture a market share beyond that maximum profit. This free rider behavior, when persistent, reduces profits of the community causing unsustainable harvests that entail the collapse of businesses and ecosystems. Policy support to harvesting aggravates the excessive harvesting, while social conventions and regulations that impede the free riding foster profitable harvest that can sustain.

The economic value of nature exceeds the harvesting of bioresources because it encompasses user and non-user values referred to the 'total value assessment' (McNeely 1988). The user values are direct and indirect. The direct ones are the productive uses of nature, for instance biofuels from rapeseed. Indirect user values refer to the maintenance of those direct uses, for example, the regulatory cost related to the biofuels from rapeseed. The non-user values are divided into option, bequest, and existence values. Option values refer to the appreciation of products based on nature, for instance, the appreciation of biofuels. Bequest values address appreciations of future generations. Given the unknown preferences of future generations, maintaining availability of ecosystems and species is advocated, referred to as the

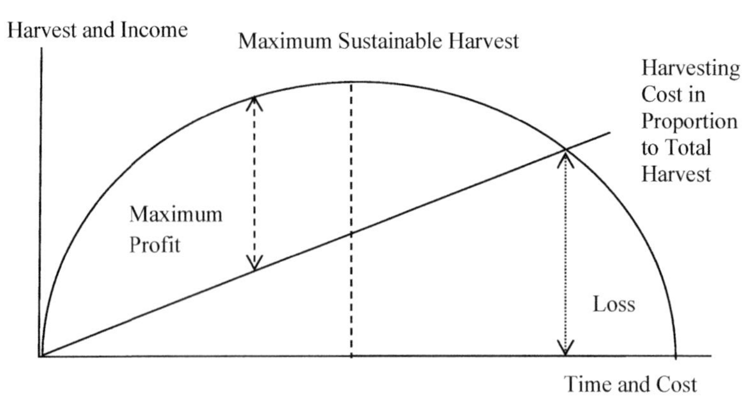

Fig. 3.1 Maximum sustainable harvest versus maximum profit

'precautionary approach'. Existence values refer to the ethical viewpoint that nature is valuable in itself, called 'eco-centric ethics' as opposed to those anthropocentric ethics focused on utilities. Eco-centric ethics are reflected in the cultural expressions of nature; for example, membership of the nature organizations.

3.6 Conclusions

Delivery of bioresources involves transformations of inputs into valuable products, which requires much know-how, as well as laborious negotiations between stakeholders that usually pursue self-interests despite the benefits of co-operative approaches. Production is also risky because it necessitates expenditure before income can be generated. Regarding the risks, the usual strategy of bioresource producers is to intensify production, measured by volume per area; but this strategy pushes the producers van bioresources in the direction of an uneconomic lock-in because the total costs of resources increase while prices of products decrease. Alternative strategies are pursued, such as the minimization of input costs, introduction of high value specialties, diversification of services, and dedicated supplies to niche markets. These strategies are possible because the customers' demands encompass qualities related to the market prices, those qualities that are appreciated but not priced as well as qualities regulated by authorities and stakeholders' negotiations. Income that can be generated from investments in bioresources can be derived from the present values of private uses and from common goods. Incomes from private uses are discounted at an interest rate higher than the considerations about common goods that are deficiently priced. Regarding these deficiencies, the decision making about bioresources should consider the user and non-user values of nature.

References

Boardman A et al (2010) Cost-benefit analysis. Pearson

Brent RJ (2006) Applied cost-benefit analysis. Edward Elgar, Cheltenham

Campbell H, Brown R (2003) Benefit—cost Analysis. Cambridge University Press, Melbourne

Constanza R, d'Arge R, de Groot R, Farberk S, Grasso M, Hannon B, Limburg K, Naeem S, O'Neill RO, Paruelo J, Raskin RG, Sutton P, van den Belt M (1997) The value of the world's ecosystem services and natural capital. Nature 387(15):253–260

Devinney TM, Auger P, Eckhardt GM (2010) The myth of the ethical consumer, 1st edn. Cambridge University Press, Cambridge

Ferguson N (2009) The ascent of money, 2nd edn. Penguin Books, London

Kahnemann D, Knetsch JL (1992) Valuing goods, the purchase of moral satisfaction. J Environ Econ Manag 22:57–70

McNeely J (1988) Economics and biological diversity, 1st edn. International Union for Conservation of Nature, Glad

Milcu AI, Hanspach J, Abson D, Fischer J (2013) Cultural ecosystem services: a literature review and prospects future research. Ecol Soc 18(3):44. http://www.ecologyandsociety.org/vol18/iss3/art44/

Millennium Ecosystem Assessment (2005) Ecosystems and human well-being: synthesis (PDF). Island Press, Washington, DC

Ostrom E (1990) Governing the commons: the evolution of institutions for collective action. Cambridge University Press, New York

Pearce DW, Turner RK (1990) Economics of natural resources and the environment. Pearson Education, Harlow

Sedláček T (2011) Economics of good and evil; the quest for economic meaning from Gilgamesh to Wall Street. Oxford University Press, New York

Simon H (2006) Altruism in economic. Am Econ Rev 83(2):156–161

Smith A (1776) Inquiry into the nature and causes of the wealth of nations, Book IV, Chapter 8.49, Edition 1986. Penguin Classics, London, p 719

Veblen T (1892) The theory of the leisure class. http://www.gutenberg.org/files/833/833-h/833-h.htm. Visited 23 Aug 2014

Zerbe RO Jr, Bellas AS (2006) A primer for benefit—cost analysis. Edward Elgar, Cheltenham

Chapter 4
Business in Bioresources

Yoram Krozer

Individuals capable of combining resources into value-added products are known as entrepreneurs. The challenge of an entrepreneur is to ensure that the cost of investments is kept below the income from sales, thus making a profit for their investments that enable continuity of its enterprise. The results of investments are inherently uncertain since expenses are made before the sales in future, and sales are guided by customer demand that is difficult for an individual entrepreneur to influence. Once the investment is made, entrepreneurs are left with little choice but to reduce costs of resources during operations and product prices for sales, which inadvertently put pressures on profits. Business models are used in preparation for those investments, meaning income generating methods. Business strategies and models for bioresources are indicated.

4.1 Business Strategies

Finding opportunities for investments that satisfy customers' demands characterizes entrepreneurial capability. Entrepreneurs are often associated with profit making, but entrepreneurial activities can also be found in research, social organizations, policy making, and other non-profit organizations. In all activities, even the most capable entrepreneurs need the assistance of specialists who can get funding, reach customers, find suppliers, and execute manufacturing and sales. The specializations enable the execution of tasks cost-effectively, also called the division of labor. Rather than involving each specialist individually, entrepreneurs often bundle such executive capabilities into an organization with internal rules and agreements, referred to as a firm, company, or an enterprise. The bundling reduces costs of transactions because

Y. Krozer (✉)
University of Twente - CSTM, Enschede and Sustainable Innovations Academy, Amsterdam, The Netherlands
e-mail: krozer@xs4all.nl

Y. Krozer and M. Narodoslawsky (eds.), *Economics of Bioresources*,
https://doi.org/10.1007/978-3-030-14618-4_4

rules prescribe execution, mitigates the risk of error due to procedures, generate larger finances and enables larger innovative efforts. Firms with similar characteristics of products are also referred to as a 'business', for instance, an energy business, food business, and so on.

4.1.1 Profitability

In the mainstream view, an entrepreneur is considered a person who is a rational profit maximizer on a market based on informed deliberations about alternatives for sales on markets. An entrepreneur, it is argued, pursues the highest profits, measured by the net present value. It implies that superior capability generates better results measured by profit. Entrepreneurs, however, have various interpretations of the profit making, in order words—about the profitability of activities, uncertainties of the future, and profits at the moment of decision making.

Some entrepreneurs pursue maximum shareholders' value within the shortest period of time, meaning the highest profit to the owners of the company in the short term. This interpretation is often observed when investors in a firms' equity are involved, for instance hedge funds. A shareholding firm can also prioritize the continuity of operations, which usually impede such aforementioned profit maximization in the short term because resources are necessary as inputs for future activities; for instance, to accrue a larger market share or innovate. Some shareholding companies focus on goodwill as the driver of profitability, such as on the impacts of operations, consumer satisfaction, or sustainability. Since many companies operate based on family capital, the consolidation of capital within the family is often pursued and prioritized rather than the dispersion of ownership across shareholders; even at the risk of lower profitability and market share. Co-operatives and stakeholder-based firms consolidate capital within broader groups of employees, participants, and sponsors, which imply that profitability is also considered in terms of social relations and environmental quality. Membership-based firms, such as pension funds or insurances, are focused on providing satisfactory services to their members while profitability is merely a condition for such services. There are also state-owned firms that pursue profitability along with the state or societal goals, as well as non-profit trusts and foundations that operate as companies but focus on their statutory goals rather than on profit. All these strategies can be successfully measured by profitability because companies that consider social and environmental issues at a lower profit in the short term are usually more trustworthy and resilient to changes in society and the stock market, as compared to those firms that aim at fast returns.

The behavioral theory of firms points out limitations in entrepreneurial decisions because rationality is bound by uncertain profits. In this viewpoint, entrepreneurs have a chance of sales on markets, but no certainty. Many models for decision making can be found, but they all deliver only a chance for profit. When operating under uncertainties about future profits, there is no single best investment decision, but a normative choice based on the expected profits of alternatives. Even if the maximum profit is pursued, expectations are based the entrepreneurial attitudes.

Three typical types of attitudes are distinguished: a risk-taking attitude (optimistic), risk-avoiding one (pessimistic), or loss-avoiding attitude (regret); intermediary positions can be found as well. The risk-taking attitude addresses investments in the most profitable alternative available on the markets referred to as the 'Maximax decision'. The risk-avoiding attitude addresses the lower profit and less risky alternatives on the markets. The decision made in favor of the highest profit among the lowest profit alternatives is called the 'Maximin decision'. The loss-avoiding attitude deliberates the opportunity costs; it means the highest profit from which is subtracted the profit of an alternative. The investment decision is in favor of the lowest opportunity cost on the markets, also called 'Minimax regret'.[1] Entrepreneurial attitudes are decisive. See Box 4.1.

Box 4.1 Decision under uncertainty An entrepreneur that deliberates investment in biogas or pellets for the household and businesses markets faces four net present values (NPV).

The risk-taking entrepreneur will favor the highest NPV across markets, which is biogas. The risk-avoiding one expects that the lower NPV is realistic and chooses the highest one among them across markets, which are pellets. If a chance of NPV is included, the expected value of profit changes.

Choice is bold	Household market	Business market	Maximax risk-taking	Maximin risk avoiding	Expected value
Biogas	24	75	**75**	24	*0.6 * 24 + 0.4* 75 = 44.4*
Pellets	65	32	65	**32**	*0.6* 65 + 0.4* 32 =* ***51.8***
Chance	*0.6*	*0.4*			

Loss-avoiding entrepreneurs estimate the opportunity cost, which is higher for biogas on the household market (65 – 24) than the cost of pellets (65 – 65), but the opportunity cost of biogas on the business market is lower (75 – 75) than that of pellets (75 – 32). The lowest total opportunity cost is in pellets.

[1]Basic introduction can be found at: https://www.youtube.com/watch?v=NQ-mYn9fPag.

Opportunity costs, choice is bold	Household market	Business market
Biogas	65 − 24 = 41	75 − 75 = 0
Pellets	65 − 65 = 0	75 − 32 = 43
Total	**41 + 0**	43 + 0

Behavioral theory shows that there are various optimal decisions beforehand (*ex ante*), though assessors can evaluate the best option after the decision making process (*ex post*). Hence, there are various 'best' entrepreneurs. In practice, the decision making is rather habitual, based on attitudes within a company often referred to as culture in organizations.

4.1.2 Innovation Process

Innovations add an element of uncertainty in decision making because the uncertain realization of a novel product adds to uncertainty in future demands. There are differing opinions about how to interpret innovating activities.

In the mainstream viewpoint, innovating is a demand-driven activity. It means that entrepreneurs act when they foresee demands for novel products as well as low competition in the future markets; for instance, it could be the development of bio-kerosene for aircrafts with respect to demands triggered by policies on climate change and only a few competing developers. Entrepreneurs innovate when they expect large profits, because investments in product development are large and profits uncertain considering that development can fail and market demands change after several years of development. Innovating is conceived as a process of scoping alternatives toward a targeted, specific innovation called 'funnel'. Initially, ideas about novel products for future markets called 'product-market combinations' are forged by selecting from the best ideas for research. This is then followed by the development of a novelty, called an 'invention'. After a successful research and development phase (R&D), the novelty is demonstrated to clients, produced in pilot mode and then scaled up for sales in the business phase; maturity in the product development toward sales is sometimes subdivided into ten or more technology readiness levels.

In that funnel, failure can occur at every phase. The failures are, for instance, researchers find errors in the selected ideas, developers encounter deficiencies in research, demonstration tests indicate customer dissatisfaction, pilots deliver poor quality, and sales can be disappointing when demands change. In reality, only a few percent of all ideas turn into inventions and a few percent of all patented inventions are actualized into profitable innovation. That said, the chances of success increases with the progress of every phase because know-how is generated and the investment costs increase; the investment multiplied by the chance of success indicates the investor's risk. When policies subsidize a high level of technology readiness, they can reduce uncertainties and spend much money but this does not reduce the risk and when they support a low level they spend less money but the chance of loss is high.

Investors aim to minimize their risk based on the opinions of experts and customers. However, experts are biased by their specializations and interests, and customers can rarely assess the value of novelties before using them. Another approach for risk reduction is screening the track record. That screening refers to the market performance, such as consumer satisfaction in the past, and to the innovator's resources, for instance, scale and quality of its R&D. A scant track record is perceived as risky, nonetheless, the contrary is often observed: new firms and new managers in vested firms often generate more successful innovations than vested ones, mainly because their motivation is high (Kaplan et al. 2009). In practice, firms' risks are reduced primarily through subsidy, copying market-ready novelties, and the spread of investment across many novelties.

Demand-driven perspective on the process of innovation mentioned above is disputed because demands are intangible at the start of an innovation process; also, it is impossible to foresee demands, the cost of product development and sales in the far future. It is therefore argued that innovations do not stem from demand but due to deficiencies perceived by some entrepreneurs and imperfect information on markets. In the heterodox non-conventional view, the innovating process is driven by entrepreneurs capable of scanning deficiencies on markets with the objective of defining and realizing a novelty with imperfect information, before competitors can achieve that. The deficiencies indicate market opportunities for a novelty. Rather than market demands, the capabilities to identify deficiencies, turn them into firms' opportunities and possibilities to generate the executives' skills are considered to drive innovation (Kirzner 1997).

The idea of a funnel is also disputed because various alternatives are tested, by trial and error, in rounds of experiments rather than striving for the targeted product-market combination. Know-how combined with social skills for execution is decisive for innovations. The alternatives are selected on the basis of these individual capabilities of entrepreneurs called 'effectuation' (Sarasvathy 2001). As per this behavioral viewpoint on innovating, research and development is perceived as the generation of the knowledge that is already available among suppliers, investors, experts, and other stakeholders, referred to the 'knowledge spillover'. Herewith, knowledge spillovers are comprehended as a key resource for innovations while business development and the high density of stakeholders' know-how foster knowledge spillovers, for instance cities with many students.

4.2 Biouresources Based Business

Theorizing of business strategies is useful for assessing opportunities in the markets of bioresources. The bioresources market is largely based on the production and consumption of crops for food, fodder, fibers, and residues, as well as wood products, and meat. All market data are mere guestimates because statistical data about the products and markets are deficient. Empirical data are focused on the reality in the European Union.

4.2.1 Bio-Based Products

The global production of bioresources approximates 14,768 million tons a year; about half of which comprise food crops. The European Union, herewith, is a large producer with an output of about 1965 million tons of bioresources, accounting for 13% of the global production from a region that constitutes about 7% of the global population. Bioresources markets in the European Union are valued at USD 2357 billion a year with nearly 22 million jobs in the production, processing, and sale of bioresources (Scarlat et al. 2015).

The total market of bioresources can be divided into a few categories. The lowest price category is bioenergy. The bioenergy market in the European Union bases itself on 390 million tons of biomass delivering biofuels equivalent to about 123 million tons of oil. This is nearly 63% of global renewable energy consumption covering about 18% of the global total energy consumption in 2015; all of these data have been sourced from European statistical data (Eurostat). The biofuel market constitutes about USD 60 billion, next to a large local use of biomass for energy.

Industrial products are smaller by scale but higher in price, and more diversified than bioenergy. These cover forestry and wood products; as well as agro-industrial products based on starch, sugars, bio-based chemicals and bioplastic; lubricants, solvents, surfactants, and other biochemicals. In the European Union, the wooden products cover nearly USD 500 billion a year, agro-industrial ones about USD 230 billion and biochemicals about USD 50 billion a year. The prices are usually around USD 1000 per ton biomass (Biddy et al. 2016; van Haveren et al. 2008).

The largest and most diversified bioresources market is that of the nutrition of people and cattle. It constitutes more than USD 1000 billion a year and uses about one billion tons biomass in the European Union. Nearly as large are the diversified markets of high value products of personal care and medicine; however, the share of bioresources in it remains unknown. The prices per ton are often many times higher than food prices, up to USD 25,000 per ton of flagrance (Bio-based Industries 2016).

4.2.2 Bioresources for Energy

Bioresources also act as a sink for pollution. When bioresources grow, they absorb carbon dioxide, and when organic matter decomposes without oxygen (*anaerobe*), methane is generated. Methane can replace fuel. This replacement is profitable when the organic matter is concentrated, for instance in the waste streams from farms, breweries, penicillin production, and sludge in wastewater treatment facilities. Additional decomposition of organic matter in the air with oxygen (*aerobe*) releases carbon dioxide and water but the releases can be prevented by the use of plants because they absorb carbon dioxide during growth; the plants also provide some benefits such as fuel, fodder, and amenities. A major cost factor of the use of plants is space because the decomposition takes anywhere from several days to weeks. Carbon is a building material of plants. Carbon storage in bioresources is much debated with respect to mitigation of climate change because carbon dioxide is the main greenhouse gas, and therefore further elaborated hereunder.

Carbon storage due to the production of biofuels and biochemical plants is referred to as the '1st generation of biofuels'. Plants can be annual crops, such as rapeseed for biodiesel; and perennial crops, for instance, willows for fuel chips and fuel pellets. In addition, local production of biofuels and biochemicals generates local income when money circulates within the community rather than being spent on internationally traded mineral fuels. Since efforts in innovations are not needed and low-productive areas can be used at a profit, it can be considered a risk-avoiding business strategy (Bentsen and Felby 2012; de Wit et al. 2011; Popp et al. 2014). The main drawbacks are competition with higher value products for land, as well as the degradation of ecosystems when agricultural monocultures exhaust and pollute the soil and water with fertilizers and pesticides whose production requires energy that results in carbon dioxide emissions. In effect, social costs of 1st generation biofuels can exceed the benefits of carbon storage. Hence, that strategy is disputed.

Biofuels can also be generated from organic waste, referred to as the '2nd generation of bioenergy'. Given that organic waste involves disposal costs, the production of bioenergy can be beneficial when waste disposal is prevented and technologies for processing waste into biofuels are available. Table 4.1 shows the annual volume of organic waste in the European Union, and potential biofuels from the period

Table 4.1 Organic waste in Europe in 1000 tons per year, own estimates based on various reports

Sectors	Supply	Reuse	Bioresource	GJ/Ton	PJ a year	Million t.o.e.
Households	87,178	20,865	66,313	10	663	15.8
Paper	92,300	66,918	25,383	11.5	292	7.0
Sludge	10,177	4556	5621	4	22	0.5
Manure	1389,258	694,630	694,630	1.65	1146	27.4
Wood	436,843	30,688	34,839	8	279	6.7
Agriculture	367,000	244,000	123,000	8	984	23.5
Food industry	37,000	–	37,000	8	296	7.1
EU total	2,419,756	1,432,973	987,785		3682	88

The biofuels from biomass residues is: $PJ = \sum_{n=1}^{Q} (Q_{tot} - Q_{use}) \cdot k_e$ when GigaJoule (GJ) and PetaJoule (PJ) are energy units, Q_{tot} is biomass production and consumption per sector, Q_{use} is biomass reused and k_e is converted biomass in energy

2012–2015. Rows show the largest organic waste streams, and columns the scale of supplies, present reuse and residues for biofuels, as well as the energy content per ton of the residue and energy equivalent of biofuels in a year. The total organic waste is about 640 million tons a year. The biofuels based on those waste streams can deliver an equivalent of 74 million tons of oil per year, which substitutes about 6% of all energy consumption in the European Union; and reduces 160 million tons CO_2 using 2.20 tons CO_2 per ton oil equivalent of natural gas.[2] The utilization of organic waste for biofuels can cover about half of the targeted 195 million tons oil equivalent by the European Commission; but it is risk-taking because logistics from scattered sources of waste, and processing mixed qualities of organic matter must be arranged.

Pilot production with algae and yeast refined into biochemicals is referred to as the '3rd generation of biofuels'. Such a type of biorefining delivers bulky biochemicals for plastics in the price range of USD 700 per ton of bio-methanol to USD 4500 per ton of bio-xylitol or even higher priced chemical specialties. Substitutions of the persistent, mineral oil-based plastics for biodegradable plastics are costly but demanded with regard to social concerns about litter. These concerns drive up the production of biodegradable plastics in the European Union to 6.6 million tons in 2016; the sales of biodegradable plastic are USD 15 billion market value that grows by 4%.[3] The market of sawn wood, however, stagnates because more plastic and metal is used; but innovations in bioresources can range from insulation, panels, filling materials, geotextiles, and other fiber products that meet the social demand for lowered health risks from foams, binder, and other chemicals. The biorefining of a plant can deliver several biobased products that rival products based on mineral resources. Table 4.2 shows an example of possible bulk uses of the common reed (*Phragmites australis*) with biorefining.

Biorefineries based on the seasonal and regional variation of bioresources are an innovative concept of downscaling technologies which emerge in foods, dyes, and

Table 4.2 Possible bulk uses of reed

	Energy pellet €/kg	Insulation panel €/m^2	Roof thatch €/m^2	Fodder €/kg	Downy filling €/kg
Plant	Residue	Stem	Stem	Stem	Flower
Processing cost	0.06	2.7	2.3	0.05	N.A.
Product prices	0.08	0.6	2.4	0.07	85
Periods	Winter	Winter	Winter	Summer	Summer
Challenges	Logistics	Weaving	Handwork	Logistics	Demands

[2]http://www.iea.org/bookshop/729-CO2_Emission_from_Fuel_Combustion. Accessed 26-7-2017.

[3]https://european-biotechnology.com/up-to-date/latest-news/news/bioplastics-market-growth-drops.html.

other products. This downscaling is challenging from the technological perspective but can be attractive from the societal viewpoint when such innovations generate backward linkages toward the diversification in agricultural production, entailing products related to specific ecosystems. Regarding the growing demand for dedicated, high value products, revenues from such specialties can compensate for higher costs associated with a smaller scale. The biorefineries in tune with regional conditions can turnaround capital-intensive economies of scale of the bulky chemicals, into knowledge-intensive economies of scope for specialties.

4.2.3 Social Demands

The innovative approaches to bioresources depend on the social demands for emission reduction, healthy foods, fair social relations, and other ethical attributes in products. Assessment of such social demands is useful for innovation. Social demands can be assessed with the use of contingent valuation, but opinions usually overestimate purchases. An alternative is to estimate the costs of compliance with the social demands as a baseline for innovations that deliver cheaper solutions. Figure 4.1 shows this concept of life cycle costing focused on social demands for emission reduction, which also applies to the demands for other ethical attributes (Krozer 2008).

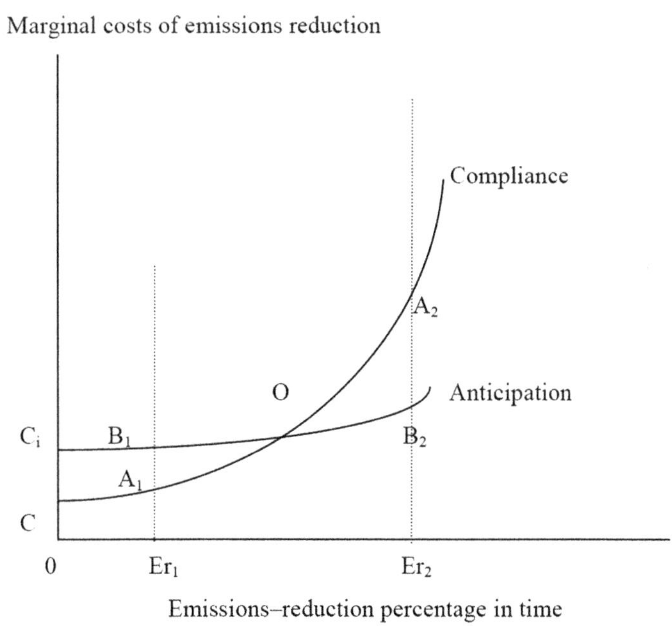

Fig. 4.1 Emissions reduction percentage in time

The demanded emissions reduction percentage, Er, is shown horizontally. The expected marginal costs of emissions reduction, Cr, are shown vertically. A stricter demand for emissions reduction percentage is shown in the shift from the position Er_1 to the position Er_2 on the horizontal axis. The costs as functions of emissions reduction are shown by the lines A and B. The compliance cost function, A, is based on available technologies from the past. These costs increase exponentially. The anticipation cost function, B, shows that the costs of change-over must be made before an investment at Er_2; but the costs of emissions reduction do not increase much, and can even decrease in the case of beneficial activities such as renewable energy and material savings or increased sales due to better products. The marginal cost of the anticipation strategy compared with the marginal cost of the compliance strategy is higher at low demand, Er_1, but lower at strict demand, Er_2. The present value of the total costs of change-over given in the figure is represented by the area A_1OB_1. The present value of the total benefits of the innovation in comparison with the available technologies is represented by the area A_2OB_2 (representing the surplus of innovation rents). The anticipation is attractive if $A_2OB_2 - A_1OB_1 > 0$. The model implies that the anticipation strategy using innovations is usually costly with moderate demands, but can be attractive with strict environmental demands. The social demands can be approximated by policy goals for the future.

The estimation is implemented in a few steps. First, the materials flow and costs per step in the value chain are estimated based on the market prices. For instance, in case of biodiesel based on rapeseed, the value chain includes the—materials and costs of producing inputs for cultivation; cultivation of rapeseed; processing to rapeseed oil; production of rapeseed methyl ester as biodiesel; and finally, the use of that biodiesel in cars. Second, the volume of pollutants and pollution reduction targets are defined based on policy documents. These cover pollution related to fuel consumption, as well as that which is caused by fertilizers and pesticides in agriculture, water pollution in cultivation and processing, and air pollution during the use of biodiesel. Third, the unit costs of technologies for pollution reduction are collected to estimate the total costs of pollution control, which is pollution multiplied by the reduction percentage multiplied by the unit costs. The sum of the present costs in the market prices and potential costs of pollution reduction indicate the social costs of producing the fuel. The calculation procedure and results for biodiesel as compared to diesel are shown in Box 4.2.

Box 4.2 Example of life cycle costing The calculation procedure is shown schematically:

	Costs in USD	Materials and pollutants in kg	Targets %, unit cost USD/kg	Pollution control USD
Production		Pollution per step in value chain	Unit cost of the pollution reduction per step	
Processing				
Distribution				
Consumption				

Final result of the life cycle costs of biodiesel based on rapeseed methyl ester compared to diesel.

Cost is in €; 2.52 kg CO_2 per kg fuel	253 000 km is about 25 tons fuel	Production cost	Cost of use	Total cost	Tax	Total with tax
Diesel Costs	Fuel (prices)	1635		1635	8920	10,555
	Pollution control costs	755	10,875	11,630		11,630
	Total	2390	10,875	13,265	8920	22,185
Biodiesel	Fuel (prices)	6965		6965	1529	8494
	Pollution control costs	2306	10,593	12,899		12,899
	Total	9271	10,593	19.864	1529	21,393[a]
Biodiesel Benefit		−6881	282	−6.599	7.391	792
[a]Out of it, the CO_2 sink of 64 tons is						6089

Biodiesel production is costlier than diesel but its use somewhat cheaper. Diesel is taxed, not biodiesel. The biodiesel life cycle is cheaper than diesel when tax exemptions for the mitigation of climate change and far reaching pollution controls are introduced.

The assessment indicates benefits that can be generated due to the social demands for ethical attributes. The market for bioresources is enhanced by the far reaching social demands for pollution reduction expressed as liabilities, or market prices with high pollution charges.

4.3 Business Models

Private and social demands give rise to many new companies. In the European Union, for instance, about 2.5 million new companies emerge every year, but more than 2.0 million die; both numbers exclude holdings. The start of a firm does not guarantee its survival. Various methods of income generation emerge referred to as the 'business models'. Sound preparation of a business model fosters continuity.

4.3.1 Business Models

Ten strategies for an innovative business model are identified. Each strategy addresses a particular element of business and contains a few alternative operations. When those strategies are combined innumerable business models can be identified (Keeley 2013).

The first strategy refers to the profit model. It is about the method of making profits. The method can be based on premium prices, auctions, metering, and subscriptions. In bioresources, premium prices are paid for ethical attributes but grains and commodities are often traded in auctions. The income from sales of biodiesel in the gasoline station is based on metering of the volume of sales. There are also subscriptions to the volume; for example, subscriptions to deliveries of ecological foods are widespread.

The second strategy refers to networks. It is about the social relations. The collaborations with stakeholders can be based on open innovations, creating secondary markets with consumers, as well as franchising for brand and quality. In bioresources, the genomics project is an example of open innovations though many follow-up innovations are licensed. A secondary market is, for instance, the reusing of residual wood for fuel. Restaurant chains are often based on a franchise model, which means the stores are privately owned but jointly supplied and promoted.

The third strategy refers to the internal structure. It is about how the firm's structure facilitates income generation. The facilitation can be based on incentives, standardization, or knowledge centers that support sales. In bioresources, the incentives are for instance, the accounts of carbon dioxide storage in forestry as presented by some airlines for compensation of pollution caused by flights. Qualities of commodities are often standardized because this reduces the costs of quality control and transactions, for instance percent of methane in biogas. The centers of knowledge are usually found on dedicated markets, for example, markets in herbal medicine.

The fourth strategy refers to processes within a firm. It is about changes in production which cover lean production, process standardization, and predicting analysis.

In bioresources, lean production can be achieved by the reuse of organic residues because nearly all residues are reusable for a valuable product. The standardization of processes is laborious because the volumes and qualities of bioresources fluctuate with time and space but genetically homogenous varieties are pursued. The steering of processes guided by predictive analyses, therefore, can resolve standardization issues, for example, predicting products of biorefinery based on the characteristics of biomass.

The fifth strategy refers to the product. It is about the subject of sales. Focus on the performance of products encompasses simplification, sustainability, and customization of products. In bioresources, wood pellets are an example of simplification in handling biofuels. Sustainability is encouraged, for instance, in biodegradable plastics as it reduces pollution in comparison with mineral-based plastics. The customization of bioresources is popular when crafts make use of biodiversity, for example, delivery of artistic or artisanal wood products.

The sixth strategy refers to the product system. It is about how products are delivered. It can range from product bundling, innovation platforms, extensions of products, product–service combinations to complementary offerings. Products that are bundled are, for example, ecological foods packed as a delivery basket. The innovation platforms are popularized, for instance, as exchanges of the cooking recipes. Extensions of products are, for instance, repairs of furniture. The product—services systems of bioresources are deliveries of ready meals.

The seventh strategy addresses services. It is about deliveries of non-tangibles and labor intensive activities. Services refer to enhancement or product uses, maintenance plans, customer support, information and education, warranties, and guarantees. Product uses are enhanced when services explain complex products, for instance support of personal care products. Maintenance of nature resorts, parks, and other common land are typical service strategies that generate income. Customer support in personal care and medicine is conventional, whereas cultural expressions of nature are popular in education. The dedicated markets of bioresources largely depend on reliable warranties and guarantees expressed by labels.

The eighth strategy refers to channels. It is about the media used to reach out to customers, such as flagship stores, pop-up stores, direct and indirect sales, and multilevel marketing. The flagships of ecological and fair trade products are widespread in foods, textiles, and toiletries. Pop-up stores with food on the market, as well as during events are popular. Most biochemicals and biofuels are sold based on direct marketing, while indirect marketing is linked to institutions that are credible in nature, for instance products supported by a social organization. Multilevel marketing is linked to outstanding qualities, such as regional wines.

The ninth strategy refers to branding. It is about the methods of marketing, which encompasses offering extensions, creating big ideas, setting values, and branding of components. Branding of the bioresource can overcome the hurdle of higher prices. Extensions are, for example, vouchers for the sales of organic foods. Ethical attributes are generally based on the creation of ideas, for instance, a positive idea about vegetarianism and set of values are related to veganism. The branding of components is often linked to approval by tests of social organizations.

The tenth strategy refers to customers' engagement. It is about involving people through the use of social media, creating simplicity in complex areas, trusted partnerships, packaging of customer experience, and so on. Social media is widely used for voluntary activities related to bioresources, for instance, the cleanup of animals after disasters. Experts and artists are involved in explaining the damages against nature. Partnerships and trust play an active role in the protection of nature and co-creation of products with ethical attributes, and a customers' experience is enhanced in nature parks.

Unique business models are created when a few strategies are combined, of which a few examples are elucidated. The redesign of furniture in combination with take-back and refurbishment services involves a design for recycling (product performance strategy) and reverse logistics of the discharged furniture (product system strategy). A more complex example is biological pest control. A producer, herewith, added non-cure no-pay services (profit model strategy), in combination with a conference center for customers (structure strategy) and educating the customer about the use of such controls (services strategy). The last example is about a theater festival that combines several strategies—the unique qualities of an island (network strategy), with attracting customers as eco-innovators (channels strategy), branding of local culture (branding strategy) and involving volunteers in the organization of the festival itself (customers' engagement strategies). Such strategies elevate the scope of innovations beyond changes in processes and products, while combinations enable the definition of a unique business model.

4.3.2 A Start-up

The business model of a company can be developed in several steps (Osterwalder and Pigneur 2010). The steps and main issues are presented using the example of reed for insulation panels. The main steps and basic data are shown in Fig. 4.2.

The development of a business model starts with a value proposition, which means the prime objective of the future business. In consideration of the value proposition, questions should be answered such as: what products and services are we offering to the target customer? How do these products and services satisfy customer needs? The value proposition should be justified by tangible deficiencies on markets or compulsory social demands for improvements. The justification for development of reed is the substitution of chemical insulation which is cheap but poses health risks and is non-renewable. Reed can be a renewable resource for durable insulation panels that enable energy saving in houses, whereas the residues from the reed plants are reusable for biofuels.

Then, consumer segments are addressed by answering questions such as for whom are we creating the value, and in which country or region? Is it a mass or niche market segment, or a multi-sided platform? Reed for insulation panels cater to construction companies aiming to meet the demands for insulation in the ecological housing market in Europe. Some market niches accept higher prices of the bioresource-based

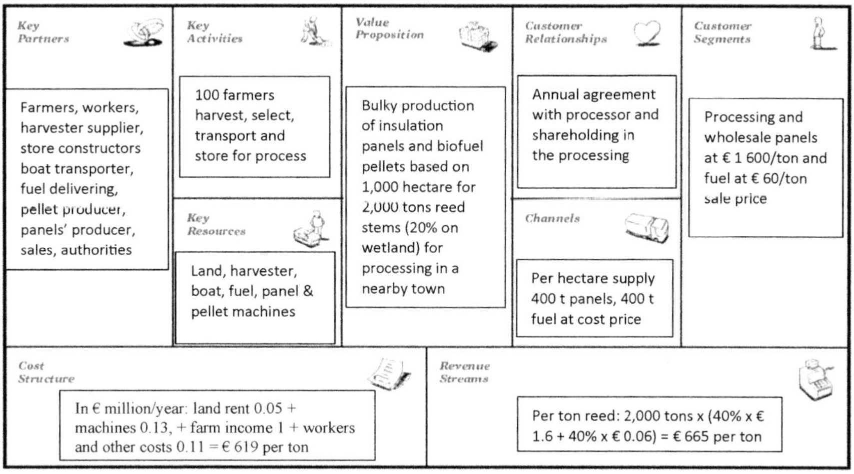

Fig. 4.2 A business model for reed-based insulation panels

insulations because they are renewable and recyclable. The reed residues can be utilized locally at a low price.

Thereafter, the channels of delivery to the customers are assessed based on answers to questions such as how can we reach out to the customer? What channels are currently used? What channels are the most cost-effective? The types of channels are usually: sales force, web sales, own stores, partner stores, and wholesale. The sales of reed-based panels can be directly delivered to the chosen construction companies (sales force) because the prior experience with sales in this case are found to be cost-effective. Web sales or wholesale are also feasible, possibly even cheaper; but services are less customer specific.

A customer relation is defined by answering questions such as what type of relationship does the customer expect? How costly and valuable are these relationships? What are incentives for purchases? How loyal do we expect our customers to be? Customer relations consist of professional assistance, dedicated personal assistance, self-service, automated services, communities, and co-creation. The reed panels for insulation can be delivered to customers based on contracted qualities and quantities of suppliers that operate on the eco-markets; this can include professional assistance. The customer relation can be supported with incentives, for example, a travel voucher to nature areas where reed is sourced.

Once all these elements are assessed, revenues can be estimated based on answering questions such as: which values are customers willing to pay for, and how do they prefer to pay—fixed or dynamic? There are many variants of pricing. Fixed price models involve listing prices. They are dependent on volume, features, and the customer. Dynamic price models are based on auctions, negotiations, and yield. Many models for generating revenues can be found, some examples: licensing, advertising, e-commerce, commission, lease, product sales, and fee for use, fermium, subscription, and mark-up. In the case of reed, a fixed price for the product is envisaged to encourage transparency with customers.

After the revenue possibilities are estimated, the resources are assessed. This is necessary in order to answer questions such as what key resources (assets) do our value propositions require? Is it research and development, production, services, delivery, marketing, distribution, or customer relationships? What unique assets will make it difficult for others to copy our solution indicated by the value control points? The key resources can be physical assets such as manufacturing, information, and communication technologies or logistics. They can also be intellectual property, for instance, copyrights and brands, clinical data, patents, and human resources such as sales and marketing, software engineers, scientists. In the case of reed, the key unique resource refers to the ecological products for insulation, in combination with guided visits to the nature reserve.

The next step addresses activities that deliver resources. It is based on answering questions such as which key activities do our value propositions require? The activities refer to research and development, clinical trials, supply chain management, production and materials, services and helpdesk, deliveries, marketing, sales, and particular add-in technologies. The reed-based insulation panels should be demonstrated to potential customers prior to conducting the assessment of all required activities for their manufacturing.

Then, key partners are assessed based on their responses to questions such as what key activities will be performed by us and by partners (or suppliers)? What key resources or assets are we acquiring from partners? Who will be our key suppliers and strategic partners? The motivation for partnership can be cost optimization, reduction of risks, and acquisition of resources or assets. Since the reed products are still at an early stage, the key activity is product development and demonstration to the customers. The delivery of reed plants must be done by farmers in reed-rich areas as this would enable sourcing of resources at a low cost.

Finally, the cost structure is determined by answering questions such as what are the important costs inherent to our business model? Which key resources and activities are the most expensive? An important issue is whether the business is cost driven or value driven. The former refers to the leanest cost and low price proposition entailing maximum automation and extensive outsourcing. The latter is focused on establishing a premium value proposition, for example outstanding quality in the sense of excellent functional performance of products and high ethical values of supplies. The business of reed-based insulation panels is value driven because it is focused on a premium value proposition at the lowest possible cost.

In theory, when the stream of revenues exceeds the stream of costs, the value proposition can be considered as feasible. In practice, numerous specifications and adaptations are made as part of assessments until the proposition is attractive to customers, suppliers, and partners. Preparation of the value proposition can take several hours or days; of the customer segments and key resources—several days to weeks; and finally, preparing the costs and revenues could take several weeks to months to complete. Moreover, adaptations ensue concurrently with business as experience is gained.

4.4 Conclusions

Businesses in bioresources encompass various strategies when they pursue profitability, which depend as much on the attitudes of entrepreneurs and culture in companies as they do on rational deliberation. In bioenergy, businesses address the processing of various plants into biofuels as well as the more laborious processing of different bio-residues up to complex, innovative biorefineries for bulk chemicals and specialties. Demands for ethical attributes of bioresources are essential for sales because they can compensate for costlier supplies based on bioresources, as opposed to mineral resources. Feasibility of supplies that take into consideration various ethical attributes can be indicated by life cycle costing. In bioresources, many start-ups are found and various business models are employed for the development of the start-ups and vested companies in the processing of bioresources. There are many ways forward for entrepreneurship in bioresources, which forge an optimistic perspective for business in bioresources.

References

Bentsen NS, Felby C (2012) Biomass for energy in the European Union—a review of bioenergy resource assessments. Biotechnol Biofuels 5:25

Biddy MJ, Scarlat C, Kinchin C (2016) A market assessment of bioproducts with near-term potential. National Renewable Energy Laboratory, Denver

Bio-based Industries (2016) A public–private partnership on bio-based industries, bio-based for growth, p 12. http://www.biobasedeconomy.nl/wp-content/uploads/2012/07/Bio-Based-Industries-PPP-Vision-doc.pdf. Accessed 5 Sept 2016

de Wit M, Londo M, Faaij A (2011) Productivity developments in european agriculture: relations to and opportunities for biomass production. Renew Sustain Energy Rev 15:2397–2412

Kaplan SN, Sensoy BA, Strömberg P (2009) Should investors bet on the jockey or the horse? Evidence from the evolution of the firms from early business plan to Public companies. J Finance LXIV(1):75–115

Keeley L (2013) ten types of innovation. Wiley, New Jersey

Kirzner I (1997) Entrepreneurial discovery and the competitive market process: an Austrian approach. J Econ Lit 35(1):60–85

Krozer Y (2008) Innovations and the environment. Springer, London

Osterwalder A, Pigneur Y (2010) Business model generation. Wiley, New Jersey

Popp J, Lakner Z, Harangi-Rákos M, Fári M (2014) The effect of bioenergy expansion: food, energy, and environment. Renew Sustain Energy Rev 32:559–578

Sarasvathy S (2001) Causation and effectuation: toward a theoretical shift from economic inevitability to entrepreneurial contingency. Acad Manag Rev 26(2):243–263

Scarlat N, Dallemand J-F, Monforti-Ferrario F, Nita V (2015) The role of biomass and bioenergy in a future bioeconomy: policies and facts. Environ Dev 15:3–34

van Haveren J, Scott EL, Sanders J (2008) Bulk chemicals from biomass, biofuels. Bioprod Bioref 2:41–57

Chapter 5
Bioresources and Technologies

Michael Narodoslawsky

The term bioresources represent a vast diversity of materials with an equally vast diversity of origins and properties. Common to all bioresources is that they ultimately utilize nature's ability to generate complex molecules from carbon dioxide, water and nutrients with the help of solar irradiation by the process of photosynthesis. Herewith, human impact on total natural biomass production is considerable; either by directly using biomass for societal purposes or the impact on land use change as a result of economic and societal activities like agriculture or settlement development.

5.1 Diversity of Bioresources

The ultimate origin of bioresources in the process of photosynthesis already delineates some important properties that have a profound impact on the structure of technological and economic utilization of these materials:

- *Bioresources require other basic resources for their generation*: Area to capture solar irradiation, carbon dioxide, water and nutrients are basic resources without which the primary production of plant biomass is not possible, as it forms the basis for all economically utilized bioresources.
- *Bioresources allow society to latch into natural material cycles of carbon, water and nutrients*: Using materials that form vast natural cycles critically reduces environmental impacts, in particular—climate change. The use of bioresources allows us the beneficial services of carbon containing materials, without risking the overburdening of the atmosphere or oceans with carbon extracted from natural long-term storages like fossil resources.

M. Narodoslawsky (✉)
Institute for Process and Particle Engineering, Graz University of Technology, Graz, Austria
e-mail: narodoslawsky@tugraz.at

Y. Krozer and M. Narodoslawsky (eds.), *Economics of Bioresources*,
https://doi.org/10.1007/978-3-030-14618-4_5

- *Photosynthesis is an infinite process as long as solar irradiation and the natural cycles of carbon dioxide, water and nutrients are active*: This makes bioresources 'renewable', meaning that they are part of perpetual global cycles rather than emanating from exhaustible reservoirs.
- *Although the photosynthesis of bioresources may go on infinitely, the basic resource area (and on a local level possibly water and nutrients) are limited*: This leads to a seemingly contradictory property of 'limited infinity' of bioresources. For while they are infinite in temporal terms, bioresources are limited in their annual yield.

Plant biomass from cultivated agricultural fields, grassland and forests as well as marine biomass harvested from aquacultures or fisheries are the starting point for all possible bioresource utilizations. The large varieties of resources as well as wide range of ingredients they contain make bioresources interesting to many sectors.

In fact, the most important role bioresources play for society is nutrition. This applies in particular to nutrition secured—directly or indirectly—from crops, grassland and fisheries. A large part of crops as well as most grasses, however, are not used directly as food but serve as feed for animals, which in turn provide meat and animal products like eggs and milk. Crops as well as slaughtered animals and animal products are then further processed, providing the bread, noodles, meat and milk products that we consume as our daily food. At any given step in this path from harvest in the field to being served as a meal on the table—that is the food value chain—harvest residues, by-products and wastes are generated. Although these materials are (usually) unfit for human consumption, they may be interesting resources for industrial and/or energetic use. These are *secondary bioresources*; example: straw (a harvest residue) that may be used for energy generation, as a building material or feed for industrial processes that generate chemicals, bioplastics or biofuels. Another example is manure from animal breeding that is a valuable feed for digesters, producing biogas that may be used as biofuel, or for generating heat and electricity in a Combined Heat and Power (CHP) plant.

Unlike crops from fields and grass, wood from forests cannot contribute to human nutrition. Woody biomass therefore is exclusively utilized in a non-food value chain. This chain either results in material products (like paper, construction timber and chemicals) or generates energy. Again, at every step in the chain, from felling trees in a forest to the final product, by-products are generated that may form secondary bioresources for utilization in other sectors.

Finally, products from food as well as non-food value chains are consumed, resulting again in material residues like human excrements, used vegetable oil, municipal biowaste, used construction wood or used paper. These *tertiary bioresources* may again serve as resources to generate material products (e.g. biodiesel from used vegetable oil, compost from municipal biowaste) and energy services (e.g. biogas from digesters using human excrements as feed).

As a rule of optimal bioresource utilization, all by-product and waste flows that may not serve as an input to other value generating processes should be conditioned in a way that allows their re-integration into natural cycles. This implies that gaseous emissions are treated so they may be safely released into the atmosphere; liquid and solid residues conditioned so they may be safely returned to fields, pastures and forests, thus closing the nutrient cycles and retaining the fertility of the basic resource that is fertile land.

5.1.1 Bioresource Use and Competition

The limitation of the basic resource that is fertile land results in an inherent competition for land use in providing different bioresources. Besides competition among various bioresources for land, there is also a competition between land for ecosystem services (e.g. rain forests and wetlands that stabilize global climate) and human appropriation for the economic use of bioresources. A vivid example of this competition is the ongoing replacement of tropical rain forests to make room for cash crop plantations, e.g. deforestation for the production of palm oil or soy bean.

The exceptional versatility of bioresources causes many sectors to compete for them. This competition is fiercest for crops growing on high-quality agricultural soil. They may be used as food, feed for animals or feedstock for industrial processes generating chemicals and biofuels. This competition also puts pressure on other forms of land use like pastures, wetlands and forests that are transformed into more profitable fields for cash crop cultivation.

This transformation engenders—in many cases—a reduction in eco services like biodiversity or climate stabilization linked to wild or less intensively cultivated land. Currently, grassland is used predominantly as forage and supports the breeding of ruminants. Emerging technologies like the *Green Biorefinery* (Kamm et al. 2010; Ecker et al. 2012), however, open new pathways to utilize grass industrially, thus also increasing competition for this resource. Wood from forests is faced with intense competition. In this case, however, the food sector is not a competitor for this resource. The competition here is between vested industries (namely pulp and paper) with construction timber, the energy sector as well as emerging industrial utilization of chemical and biotechnological pathways. It is also worth mentioning that forests usually have a higher potential to provide eco services such as climate stabilization or water retention, when compared to fields and grasslands (Hildmann 1999).

Under these circumstances of strong competition for primary bioresources, secondary—and to a certain degree tertiary—bioresources offer attractive alternatives in particular for energy and industrial use. Although usually lower quality, current energy, chemical and biochemical conversion technologies are capable of utilizing such feedstocks. The structure of technological utilization from resource to final product, the *value chain of bioresources*, depends on the type and qualities of the resource at its root. The following sub-chapters provide an overview of the critical properties of some common bioresources, alongside structural aspects of the value chains they support.

5.1.2 Primary Bioresources

Primary bioresources are those that are harvested from the ecosphere. For the consistency of the argument, in this chapter focus will be laid on terrestrial primary bioresources which are resources that grow on land. In particular, we will look at crops, grass and wood.

The process of photosynthesis that generates these bioresources is dependent on basic resources such as land (to capture the necessary solar irradiation), water and fertile soil containing nutrients; as well as the necessary soil organisms to support the growth of the bioresource in question. All these basic resources are highly dependent on and predetermined by a spatial context. This means that the natural endowment of a certain place defines not only the kind of bioresource that may be grown there, but also the quality of this resource. Besides this dependence on the natural endowment of a particular place, societal and economic factors such as the knowledge of farmers and existing infrastructure have a considerable impact on the range of possible crops and their quality.

Quite apart from their dependency on the spatial context, primary bioresources require solar irradiation for their growth. The process of photosynthesis is—in terms of energy—relatively inefficient. Maximum theoretical conversion rates of solar irradiation into bioresources are between 4.6% for C-3 plants like wheat, and 6% for C-4 plants like maize (Zhu et al. 2008). Practical conversion rates are around 50% of these theoretical values. This means that bioresources need large areas to convert water, CO_2 and nutrients into valuable biomaterial with the help of solar irradiation, albeit with low efficiency. In terms of area demand, bioresources dramatically trail other renewable energy sources (see Fig. 5.1), and fossil resources as well. In the Figure, the value for hydro power only takes the area occupied by a reservoir into account. Figure 5.1 very clearly indicates a common trait of all primary bioresources in that they need large areas for their growth and are therefore de-central resources that require considerable effort to collect.

Another common trait is rooted in the fact that photosynthesis is a process that results in an aqueous medium and inadvertently, a considerable humidity of primary bioresources. Coupled with the limited transport density of many primary bioresources, this accounts for considerable transport effort, from field to the first stage in the value chain. Table 5.1 presents properties that have a particular impact on the collection and transport of primary bioresources. It is evident that the lower the yield per hectare and content of the main ingredient—larger is the area necessary to support a certain industrial site. Larger areas, in turn, necessitate longer transport distances for collection.

Another aspect is represented by humidity and transport density. The larger the water content of fresh bioresources, higher is the expenditure on fuel for the transport of the collection vehicle to move water that does not add value. Besides, high humidity also restricts the shelf life of bioresources requiring drying in order to make the resource storable. A lower transport density and 'bulkier' material make for more of the load volume of the vehicle. If the transport density falls below the critical

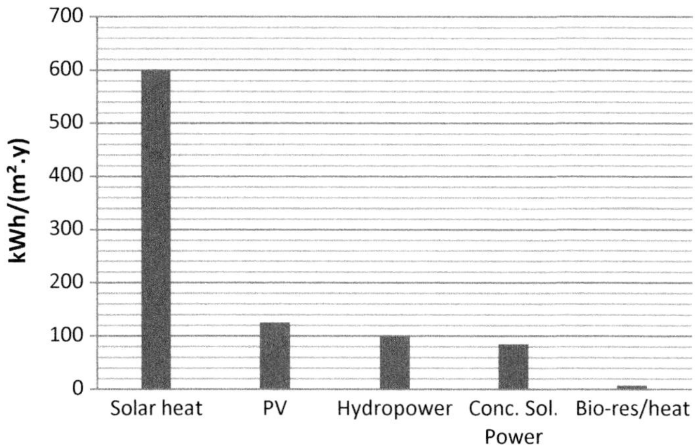

Fig. 5.1 Maximum energy yield in kWh per m^2 for some renewable energy sources converted to heat or power (Narodoslawsky 2016)

Table 5.1 Properties of selected primary bioresources [MI—Main Ingredient; FM—Fresh Mass; Density ρ_T. (t/m^3)]

Bioresource	Origin	Main ingredient (MI)	Yield [t (FM/ha.a)]	MI in Crop (% FM)	Energy Content (MJ/kg)	Moisture (% FM)	Density: ρ_T. (t/m^3)
Corn	Fields	Starch	5–15	70	15	10–25	0.7–0.8
Wheat	Fields	Starch	2.5–7	70	15	10–25	0.7–0.76
Potato	Fields	Starch	17–47	15–20	2.7[a]	70–75	0.75
Rape seed	Fields	Oil	1.3–4.2	40	24.6	6–10	0.56–06
Sun flower	Fields	Oil	1.0–2.8	48–52	24.3	6–10	0.35
Palm fruits	Plantation	Oil	4–20	20	n.a.	70	0.61–0.66
Sugar beet	Fields	Sugar	40–95	15–20	2.7[a]	75–80	0.6
Sugar cane	Plantation	Sugar/cell	40–130	10–15	7	75	0.2–0.4
Soy bean	Fields	Protein/oil	2.9	34/18	23.2	10	0.7
Grass silage	Grassland	Protein/cell	15–25	1.2–2.5/n.a.	2.6[a]	75–80	0.4–0.6
Miscanthus	Fields	Cellulose	8–17	32–42	14.2	12–20	0.12
Wood	Forest	Cellulose	3.5–6	29–32	6.8	35–42	0.75–1.2/0.3[b]
Short rotation wood	Fields/grassland	Cellulose	10–18	20–25	7	50–60	0.3[b]
Light fuel oil	Fossil	Hydrocarbon	n.a.	100	42.7	0	0.84

[a]Value represents heating value of biomethane generated by the material (as for very wet materials the heating value of fresh material is not representative)
[b]Density of chips

density of the vehicle,[1] volume becomes a limiting factor for the amount that can be transported in a single run. Consequently, this increases the number of transport runs, thus increasing the fuel consumption for collecting primary bioresources.

Finally, properties of light heating oil (LHO) are presented for comparison. The energy content (represented as heating value) of all bioresources is markedly lower than that of LHO. One reason for the lower energy content is the humidity of fresh primary bioresources, whereas LHO does not contain any humidity to speak of. The transport density of LHO is also relatively high. This comparison highlights a general logistical disadvantage of primary bioresources, in contrast to their fossil competitors.

Another common trait of many primary bioresources is their discontinuous provision. Most crops are subject to cultivation cycles with long growth and short harvesting periods, defined as seasons. Contrary to fossil resources that constantly flows from point sources, primary bioresources grow on de-central and distributed patches of land and are harvested at particular periods within a year. If they are to be utilized industrially, they must either be stored to provide a reservoir for continuously feeding industrial sites and energy provision technologies; or the utilization has to operate in campaign mode during the (short) harvesting period. The former option usually requires conditioning of the bioresources like drying (e.g. for grains) or extracting storable ingredients (e.g. oil from palm fruits), followed by storage. The latter option requires large capacity industrial sites capable of handling the harvest of a full year during the relatively short harvesting period. For the rest of the year, these sites remain idle. Several factors influence the decision to be made between these two options, such as the costs of conditioning and storage versus the costs of larger industrial installations; as well as the availability of a flexible workforce in the case of the aforementioned campaign mode of operation.

5.1.3 Secondary Bioresources

Figure 5.2 shows a value chain based on primary bioresources. Each step of this value chain may or may not be at a different location involving different actors. The figure shows that the pathway from the productive land to the final product requires a complex system of technologies as well as the interaction of a large number of actors. Managing this complex value chain and guaranteeing a balanced distribution of revenue between the different actors is a precondition for the success of the final product on the market. Energy inputs are for free, shown italic.

In most steps of this value chain, material losses as well as by-products are generated. By-products that may be utilized economically are called *secondary bioresources*. They are usually less valuable than the final products, but may considerably add to the economic viability of the value chain in entirety. In many cases, these

[1]The critical density of the vehicle is the maximum mass load divided by the maximum load volume (see also Chap. 6).

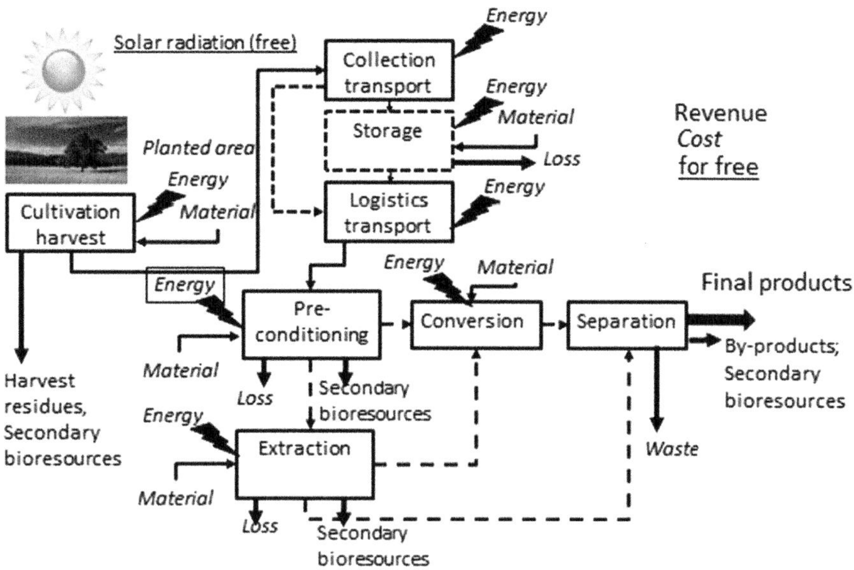

Fig. 5.2 A value chain based on primary bioresources

secondary bioresources become the root of value chains in their own right. This results in a complex, web-like topology of utilization pathways for bioresources, in which different value chains are linked via material and/or energy flows generated by using secondary bioresources. The value chain topology for bioresources reflecting the complex composition of these materials is, therefore, more complex than that of fossil resources and involves many more actors. Besides their strong inter-linkage, bioresource value chains are also contextual as the supply of bioresources depends on the regional natural endowment, as well as the existing social and economic structure. Chapter 6 will provide more insight into these systems of bioresource utilization.

Within the category of secondary bioresources, harvest residues play a special role concerning their logistics: they mirror primary bioresources in their requirement for collection logistics as they, also, are provided de-centrally. Almost by definition, harvest residues are of a lower quality than primary bioresources and in many cases have poor logistical properties too; in particular low transport densities (e.g. in the case of straw). Competition for harvest residues is, however, lesser than it is for crops. Their withdrawal from land may be limited by considerations to maintain fertility of the land: nutrients withdrawn with harvest residues may have to be replenished by additional fertilization.

All other secondary bioresources take advantage of the collection of primary bioresources they are based on. This implies that while on the one hand they emerge in considerable amount at point sources (the sites where they are generated within the primary value chain); on the other hand, they can be conditioned for transport and/or further use within the technological step that generates them. This can take the form of drying and compacting (e.g. to generate pellets from sawdust) or enriching (e.g. by steeping sugar beet pulp in molasses to increase its nutritional value as fodder). As a general rule, the logistics of secondary bioresources generated by industries along the value chain are business-to-business (B2B). This allows for the use of more efficient transport vehicles like trucks, railways and ships, in contrast to the use of (relatively inefficient) tractors that carry out most of the collection transport from agricultural land or forest.

There is always a reason to extract a secondary bioresource from the primary value chain. Secondary bioresources may concentrate the parts of a primary bioresource that are not its main ingredient (e.g. the cellulose part of sugar cane that forms the secondary bioresource *bagasse*). They may also be unavoidable couple products (e.g. manure generated in the process of rearing animals). As a rule of the thumb, secondary bioresources are of lower quality than primary bioresources and therefore their utilization demands greater technical effort. This disadvantage, however, is often outweighed by the fact that there is less competition from non-industrial sectors, in particular the food sector for secondary bioresources. Less competition from high price sectors also leads to lower prices for these resources, making them particularly attractive for the energy and chemical sector. Table 5.2 lists some properties of selected secondary bioresources, again compared to LHO as a fossil reference material.

Secondary bioresources with the main ingredient protein are typically utilized as animal feed. Materials with the main ingredient cellulose may either be used as energy carriers, or as feedstock in chemical and biochemical processes. A higher moisture content and lower transport density necessitates costlier transport. It implies that customers of such wet and 'bulky' secondary bioresources must be found closer to its site of production. There is, however, the opportunity to condition a secondary bioresource such that it can serve customers farther afield and thus command a larger market by virtue of its improved logistical properties; a good example is the case of saw dust versus wood pellets. Saw dust, on account of its low transport density and greater moisture content, is clearly a local or regional secondary bioresource. While the drying and pelletizing of wood pellets generate a secondary bioresource that can easily be transported over long distances and stored for longer durations of time; making them attractive for a large market as versatile solid energy carriers.

5.1.4 Tertiary Bioresources

Tertiary bioresources arise from the use of bioproducts by society. Examples are left-over food in municipal biowaste, used vegetable oil (UVO), sewage sludge from waste water treatment plants, waste paper, or waste wood. Their value chain, again, differs considerably from that of primary or secondary bioresources in that they emerge mainly in urban agglomerations requiring urban waste collection logistics.

The properties of tertiary bioresources are defined by the change that societal consumption imprints on bioproducts—wastewater collects the metabolic residues

Table 5.2 Properties of selected secondary bioresources [MI—Main ingredient; FM—Fresh mass; Density ρ_T. (t/m^3)]

Bioresource	Origin	Main ingredient (MI)	Yield[a] (t FM/ha.a)	MI in bioresource (% FM)	Energy Content (MJ/KG0	Moisture (% FM)	ρ_T. (t/m^3)
Harvest residues							
Wheat straw	Fields	Cellulose	1.5–4.2	36–54	15	15	0.02–0.05/ 0.05–0.11[b]
Corn cobs	Fields	Cellulose	1.2–2.2	39	10.4–12.6	15–30	0.13–0.22
Animal husbandry							
Manure	Farm	Carbo-hydr./nitrogen fert.		2–4/0.3–09	0.7[c]	60–90	1
Industrial processes							
Sugar beet pulp	Sugar industry	Protein/sugar	n.a.	14.2/7	1.73[c]	75	0.35
Rapeseed press cake	Oil industry	Protein/fat	n.a.	30–36/11–20	n.a.	11	0.56–06
DDGS	Biofuel ind.	Protein	n.a.	38	n.a.	10	0.48
Saw dust	Saw mill	Cellulose	n.a.	29–32	6.8	35–42	0.2–0.3
Wood Pellets	Saw mill	Cellulose	n.a.	55	14.4	6	0.66
Light fuel oil	Fossil	Hydrocarbons	n.a.	100	42.7	0	0.84

[a] If applicable
[b] Compacted
[c] Value represents heating value of biomethane generated by the material (as for very wet materials the heating value of fresh material is not representative)

of human food consumption; UVO is changed by the process of cooking and frying; organic municipal waste consists mainly of degraded food, chemical additives while varnishes and printing colours contaminate waste wood and waste paper. These changes reduce quality of tertiary bioresources and may complicate their technological utilization, considerably. Table 5.3 shows some properties of selected tertiary bioresources.

As is shown in Table 5.3, the quality of tertiary bioresources varies widely. While resources like UVO and waste wood are of high-quality, typical waste flows such as sewage sludge and organic municipal waste have rather high moisture content but a low content of utilizable ingredients. These latter materials are predestined for energy utilization.

However, utilization of tertiary bioresources is crucial to closing material cycles in a developed bioeconomy. These resources still contain major amounts of nutrients, in particular nitrogen and phosphor, which have been withdrawn from fertile land. Bringing these nutrients back to the land is a necessary condition to retain fertility. This, however, means that technologies particularly using low-grade tertiary biore-

Table 5.3 Properties of selected tertiary bioresources (MI—Main ingredient; FM—Fresh mass; Density ρ_T. [t/m^3])

Bioresource	Main Ingredient (MI)	MI in Bioresource [% FM]	Energy Content (MJ/kg)	Moisture [% FM]	ρ_T. [t/m^3]	Main Use
Organic municipal waste	Carbohydrates/ nitrogen fertilizers	27/1.5	2.4[a]	70	0.75	Biogas/compost
Sewage sludge	Carbohydrates/ nitrogen fertilizers	30/1.0	0.3[a]	96/65[b]	1.2	Biogas/compost
Used vegetable oil	Fat	99	39	0.001	0.92	Biodiesel
Waste wood	Cellulose	40	14.4–16	15	0.4–0.6	Energy
Light fuel oil	Hydrocarbons	100	42.7	0	0.84	

[a]Value represents the heating value of biomethane generated by the material (as for very wet materials the heating value of fresh material is not representative)
[b]De-watered

sources must be operated with a dual objective. One, to gain as much value from limited bioresources as possible; two, the simultaneous and safe re-integration of the final residues of bioproducts used by society into the ecosphere, and in particular maintaining (or even increasing) fertility by returning them to the land.

5.2 Bioresource Utilization Technologies

There are numerous technologies applied along the value chains of bioresources. The objective of this subchapter is not merely to provide a comprehensive technological description of such technologies, instead to offer a glimpse into the characteristics of common bioresource technologies and their particular role within value chains. In general, there are several roles that technologies can play within value chains. These roles strongly influence technological parameters of the aforementioned technologies, such as capacities, required input quality and mode of operation. Within a bioresource value chain, technologies may serve as *frontline technologies*, *storage technologies, production technologies* and *rear guard technologies*.

- *Frontline technologies*

  These technologies receive primary bioresources and either condition them for further use or directly generate either products or energy services from them. As they usually treat materials with poor logistical properties, these technologies tend to be realized in small units and operate close to the land on which the primary bioresources grow. Depending on the risk of degradation of the primary bioresource (e.g. if the moisture content is high), they may be forced to operate in campaign mode. A critical design factor for these technologies is their optimal integration into local/regional energy and logistical systems.

- *Storage technologies*

  These technologies serve to stabilize the value chain. They are necessary if bioresources are provided within a short (harvesting) period while the rest of the value chain operates continuously. Depending on the necessary infrastructure and level of technological complexity, they may be realized de-centrally and in small capacities (e.g. ensiling); close to the emergence of the bioresources or centrally in large capacities (e.g. controlled atmosphere storage); and often close to the utilization of the stored bioresources.

- *Production technologies*

  These technologies convert bioresources into valuable products and services. They may consist of complex process chains including pre-treatment, thermal, chemical or biochemical conversion technologies; as well as sophisticated separation and product polishing technologies. All of which are structured to provide high-quality products at the lowest possible costs, usually realized in large-scale, centralized installations.

- *Rear guard technologies*

  Rear guard technologies convert low-grade primary, secondary or tertiary bioresources into products and services while conditioning the emerging by-products in a manner that enables their safe re-integration into the biosphere. They may be realized centrally or de-centrally depending on the resources they use, properties of residues and logistics of bringing back residues to the land. Their feedstock usually consists of materials with lower quality and wider variation of composition, requiring robust technological installations.

Depending on the natural, economic and social context of the value chain, technologies may fulfil a combination of the roles described above. They may also be linked in complex, regionally adapted technology systems or *biorefineries* that are discussed in greater detail in Chap. 6.

5.2.1 Conditioning Technologies

These technologies are typical *frontline technologies* that condition bioresources for further storage, transport or treatment. They include drying, particle size reduction (e.g. shredding) and compacting (e.g. pelletizing). As a rule, they require energy input in the form of heat and power (for drying) or just power (for particle reduction and compacting). Their main objective is to improve the properties of the bioresource in order to facilitate their compliance with logistical and technological steps within the value chain (see Table 5.4). As a rule, these technologies are of low or medium complexity and they may be installed consecutively, e.g. shredding/milling wood, then drying and then pelletizing it. Depending on the logistical properties of

Table 5.4 Objectives and critical aspects of conditioning technologies

Technology	Input	Auxiliary input	Product	Purpose	Size
Drying	Grain crops, wood	Medium temperature heat, power	Dried goods	Reduction of moisture to improve storage and logistical properties	De-central for crops and wood (-chips), central for timber
Particle reduction/milling/shredding	Wood, straw, cereals	Power	Chips (wood), meal	Easier handling, increasing surface for better access to ingredients	De-central for wood, central for cereals
Compacting/pelletizing	Wood, straw	Power, binder	Bales, pellets	Increase transport density	De-central for straw, central for wood

the treated bioresource, conditioning may be realized in de-centralized, small-scale installations (e.g. drying of grain) or larger scale units integrated into industrial sites (e.g. pelletizing of wood).

5.2.2 *Storage Technologies*

Storage is necessary if the supply and demand of bioresources have different profiles over time. Moreover, the majority of primary bioresources follow the typical growth cycles of planting, growing and delivery only during the short harvesting phase—but industrial processes usually require a continuous supply of feedstock and storage is ubiquitous in bioresource value chains.

Depending on the vulnerability of the bioresource, storage technologies can range from the simple provision of space for storing goods, to technologically complex and cost intensive installations like controlled atmosphere storage or cold storage rooms. Again, more complex storage facilities tend to be larger scale and centralized, while simple storage technologies like ensiling may be applied at the farm level. There are several ways to achieve the objective of slowing the decay of bioresources. Table 5.5 provides an overview of the characteristics of different storage technologies.

Storage always implies costs, and in many cases causes loss. Minimizing the need for storage is therefore crucial in optimizing the whole value chain. Some materials (e.g. wood, sugar, some oils) are easily stored, requiring no costly technology. It is also important to analyze what is easier to store, crops or the products made from bioresources? If crops are comparatively harder to store—as in the case of sugar beets versus sugar as a product—a way to reduce costs is to operate in campaign mode which aligns the industrial utilization with the availability of a bioresource. However, in this case, the size of the industrial installation is larger; following Eq. 5.1, with A being the area planted with the crop in [ha], y the yield in [t/ha] and t_h the duration of the harvest time [in months].

Table 5.5 Characteristics of storage technologies

Technology	Effect	Auxillary input	Storage time	Side effects	Application
Drying	Reduces water content, lowers risk of biological spoilage	Heat	Months	None	Cereals, oil seeds, grass
Pasteurization/ sterilization	Kills microbes, usually applied to liquid goods	Heat, material must be sealed after treatment to prevent re-infection	Weeks to months	De-naturalization of complex compounds (e.g. proteins)	Milk
Addition of conservation agent	Establishing antimicrobial environment (e.g. by reducing pH value)	Conservation agent; may also be produced in situ (e.g. lactic acid in silage)	Months to years	Change in resource quality by conservation agent	Silage, pickles
Cooling	Reduced temperature slows microbial activity	Power, the longer the shelf life extension and lower the temperature the more power needed	Weeks to months	None	Meat, food
Controlled atmosphere (CO_2)	Reduced oxygen and partial pressure reduces microbial activity	Sealing of (large) storing space, handling	Months	None	Vegetables, fruits

$$C_{\text{camp}} = \frac{A * y * 12}{t_h} \tag{5.1}$$

Besides larger industrial sites, campaign operations also require a flexible workforce at the industrial site as they operate temporarily and sit idle for the rest of the year. Another consideration is the possibility to draw upon existing (agricultural) storage systems. Such systems exist for many cereals so that these resources, although harvested over a short time in the year, are continuously available for industrial utilization.

Taking into consideration the comparative costs of technologies listed in Table 5.5, storage with conservation agents (e.g. ensiling) produced in situ is least expensive. Thermal treatment (drying, pasteurization) is also relatively cheap, whereas controlled atmosphere storage (in particular cold storage) is expensive and only applicable for relatively high price goods, particularly in the food sector.

5.2.3 Production Technologies

These technologies form the core of value chains based on bioresources. Their objective is to convert bioresources so as to generate economic benefits in the form of marketable services or products. Different tasks may be performed within the system of production technologies, namely *extraction* of valuable thermal, chemical or biochemical ingredients; *conversion* of substances; and *separation and polishing* of marketable material products.

The size in which these technological tasks are realized varies widely, dependent on the task itself, the complexity of the technology employed and the market for the service or product. The mode of operation, too, varies. It depends specifically on the vulnerability of the feedstock, with more vulnerable feedstock (e.g. sugar beet, herbs) tending to require a campaign operation. As a rule of the thumb, value chains that provide bulk products will lead to continuously operated, large-scale production technologies.

5.2.3.1 Extraction Technologies

Extraction technologies have the objective of extracting a valuable compound from the bioresource, either for further treatment or as a product in its own right. They are always separate processes, which produce by-products that contain the part of the bioresource that is unwanted in the value chain in question. As already discussed above, these by-products may well form secondary bioresources that are utilized in a separate value chain.

Depending on the feedstock and the properties of the compound to be retrieved from the matrix (i.e. the structure of unwanted material to which the compound is bound in the bioresource), there exists a wide range of technologies to perform the extraction of the desired material.

- *Mechanical extraction/pressing*:

  This is the technology of choice if the desired compound can be separated from the matrix of the bioresource by applying mechanical stress in the form of pressure. Pressing can be scaled easily, allowing de-centralized small-scale devices that operate close to where the bioresource is grown. This reduces the effort put into transport, as collection is done within a short distance and the mass of the product retrieved from the bioresource (which has to be further transported) is less than that of the original bioresource. Besides these advantages, pressing is also relatively cheap and does not decrease the quality of the product in chemical and biological terms.

- *Extraction using extraction agents*:

  Applying an extraction agent that dissolves the desired compound away from the matrix of the bioresource is another option. The array of extraction agents are wide ranging, from hot water (e.g. used in sugar extraction) to hot salt solutions (e.g. in pulping of wood) organic solvents or steam (e.g. used in steam distillation of herbs), and supercritical gases (used to retrieve complex, high price compounds from a variety of bioresources). The choice of the extraction agent depends on the properties of the desired compound, in particular its solubility in the extraction agent. A prerequisite of an extraction agent is that it must not negatively affect the quality of the product, and that it should be easily separated from it in the subsequent steps of cleaning and polishing.

5.2.3.2 Thermal Conversion Technologies

Thermal conversion technologies subject bioresources to thermal stress in order to have them decompose. If this treatment is carried out in an oxidizing atmosphere, the conversion process provides excess heat. Many thermal conversion technologies are applied solely to generate energy from bioresources; some decompose bioresources into simple and low molecular weight compounds for further treatment. An advantage of thermal conversion technologies is their relative robustness against wide variations in quality of input material. Most thermal conversion technologies may accept low-grade primary (e.g. low quality wood chips), secondary (e.g. corn cobs) and tertiary (e.g. waste wood) bioresources.

In the case of energy provision technologies, thermal conversion generates high temperature heat. This heat can either be used directly or transformed into power and lower temperature heat employing *combined heat and power* (CHP) technologies. As a rule of thumb, larger the size of the plant, higher is the fraction of heat that can be transformed into power. This ranges from *Sterling motors* that usually show power fractions of 1–5% at thermal capacities of 15–100 kW; to *Organic Rankine Cycle* (ORC) plants with power fractions of up to 18% at capacities of up to 20 MW; and large plants (from 50 MW onwards) that may achieve between 30 and 37% power from the bioenergy input.

In terms of material by-products, thermal treatment of bioresources generates CO_2, considerable NO_x and possible fine particle emissions. The CO_2 emissions from these technologies are not regarded as contributors to climate change as they are balanced by the CO_2 fixation by the original bioresource. NO_x emissions from thermal biomass utilization are predominantly defined by the nitrogen content of the fuel and may require abatement technologies, in particular for larger scale plants close to urban settlements. Fine particles also represent an emission problem for biomass-based energy technologies, again requiring abatement technologies.

Besides gaseous emissions, thermal conversion technologies inevitably generate solid by-products, usually ashes that contain the mineral compounds from the bioresource. Re-integrating these nutrients onto the land in order to maintain fertility is an important task. Thermal conversion technologies concerning mineral nutrients are therefore always *rear guard technologies*. Nonetheless, it is important to notice that, thermal treatment is always a separation process for the mineral constituents of biomass. This often leads to concentration of unwanted contaminants (e.g. heavy metals) in particular fractions of the solid by-products. For instance, this is the case in biomass combustion wherein the ash retrieved from the grate contains low concentrations of heavy metals, whereas the fly ash shows high concentrations of these contaminants. Proper management of solid by-products from thermal conversion technologies of bioresources is therefore important.

There are several thermal conversion technologies available, generating a wide variation of services and products:

- *Combustion*

  Combustion generates only heat by fully oxidizing bioresources. The range of input material to combustion is very wide, also allowing this technology to act as a dedicated *rear guard* process in bioresource value chains. Inputs may consist of materials that are relatively dry and solid (e.g. wood chips, straw, corn cobs), and also liquid (e.g. biofuel, black liquor from pulping wood) or gaseous fuel (e.g. biogas from digestion). The more humid the input material, lower is the combustion temperature and hence the power efficiency in CHP technologies. Combustion can be realized in all capacity ranges, from a few kW to hundreds of MW.

- *Gasification*

  Gasification is the partial oxidation of bioresources, with a gasification agent, generating a gaseous product for further utilization. Usually less than 50% of the carbon contained in the bioresource will be oxidized to provide heat for the process, the rest of the biomass will generate a gas rich in carbon monoxide and hydrogen with smaller amounts of methane and higher hydrocarbons. The composition of this product gas varies depending on the gasification agent. With air as a gasification agent, the gas quality allows further combustion; whereas using steam as a gasification agent, the product synthesis gas can be used in a Fischer–Tropsch synthesis generating high-quality hydrocarbons. However, the latter option requires higher quality input and sophisticated technology in terms of gas cleaning and synthesis that is usually viable only at larger, central plants. Fischer–Tropsch synthesis is a *gateway technology* for further value chains of high-quality products from bioresources.

- *Pyrolysis*

  Pyrolysis is an endothermic decomposition of biomass, always generating solid, liquid and gaseous products. The gaseous product can be used to cover the energy demand of the process, whereas liquid and solid products may be marketed. Depending on the process parameters chosen, in specific process temperature profiles and pyrolysis times, the quality of and ratio between liquid and solid products can be varied. The liquid fraction (pyrolysis oil, up to 70% of the input material) is a low-grade biofuel that may be refined to a quality feasible for utilization in diesel engines. The solid fraction (pyrolysis charcoal, about 20% of the input) may be used as an energy carrier as well as to improve soil quality and fertility. Pyrolysis accepts an extremely wide range of input qualities and can be scaled from small to large on an industrial level. It may therefore be used as *front line technology*, transforming low-grade primary and secondary bioresources into easily transportable goods.

5.2.3.3 Biotechnical Conversion Technologies

Biotechnological conversion technologies have—arguably—the widest range of possible products that may be generated from bioresources with the help of microbes or other biocatalysts. These processes require temperatures close to or slightly above ambient temperature. Low temperatures, however, imply long conversion times requiring large equipment and considerable energy in particular power input. Biotechnical conversions require an aqueous environment, which in turn requires energy intensive or technically sophisticated separation steps to obtain the desired product. The inputs to biotechnical conversion technologies must usually fulfil strict quality standards. Sugar, either from sugar crops or derived from starch and increasingly from cellulose, is the most common carbon source for bioconversions. Besides carbon sources, bioconversions with microbes also need nitrogen sources for growth and maintenance of biocatalysts; usually a considerable cost factor.

- *Submerged bioconversion*

  This technology is the workhorse for biotechnical conversion processes. The bioconversion takes place in an aqueous solution requiring sterile conditions and relatively pure input material that must be conditioned so as to be accessible to biocatalysts. Bioconversions with life catalysts (yeasts, bacteria, etc.) will usually result in a considerable production of CO_2 and hence a loss in the carbon supplied by the bioresource. The desired product is always produced within the cells if life catalysts are used. It may either be excreted to the fermentation media (typical for low molecular weight products such as ethanol and organic acids; extracellular products) or retained in the cells (high molecular weight products like biopolymers; intracellular products). In the case of the latter, cells have to be ruptured and the product separated from the cell debris, which is usually a complex separation task. Extracellular products have to be separated from solid by-products (cells and nutrients as well as residues of the bioresource not utilized by the biocatalyst). These by-products may be valuable in themselves, like DDGS. The desired product in this case will be available in relatively low concentrations in an aqueous solution requiring considerable energy, and/or sophisticated technology for separation and polishing.

- *Solid-state bioconversion*
  Humid bioresources may directly be seeded with biocatalysts that transform part of this material into desired products, which may then be retrieved by pressing and/or extractive technologies. Due to the lower quantities of water to be handled downstream, processing may need smaller capacities than what is necessary for submerged bioconversions. This technology, however, is currently only applied to storage (e.g. ensiling) or *rear guard technologies* like composting, with a wide range of input qualities.

5.2.3.4 Chemical and Enzymatic Conversion Technologies

Besides the use of heat or life biocatalysts, chemical or enzymatic catalysts may also be applied to convert bioresources. Compared to biotechnological conversion, chemical and enzymatic conversions are usually much quicker and require smaller equipment.

- *Chemical conversion*

  Chemical conversions achieve high conversion efficiencies within seconds or minutes. However, the conditions required for the process are usually of higher pressure and temperature values than the ambient state. High conversion rates mean small equipment; while high conversion efficiencies often mean a high concentration of the desired product, thus reducing the necessary separation effort. This means that chemical conversions are usually cheap. These obvious advantages are, however, balanced by the relatively narrower range of possible products, typically simpler chemical compounds. Examples of chemically converted bioproducts are biodiesel (from oils and fats) and starch derivatives.

- *Enzymatic conversion*

  Enzymatic conversions occupy the middle ground between biotechnological conversions with life catalysts and chemical conversions. Conversion rates are considerably slower than they are for chemical conversions, ranging from minutes to hours. Conversion efficiencies may be comparable to chemical conversions. Usually, enzymatic conversions are carried out in aqueous solutions and the same arguments apply as discussed with biotechnological conversions. The product range is considerably larger than in the case of chemical conversions, including complex and valuable products. A drawback, however, is the considerable cost of enzymes, which cannot be reclaimed easily.

5.2.3.5 Separation Technologies

Separation technologies may be applied at different stages in the production process. They are, however, always present at the end of the value chain when the desired product has to be separated from other processed material and cleaned (polished) so as to obtain marketable quality. By definition, separation technologies always produce by-products which are part of the input flow that is separated from the desired product. The objective of separation technologies is to generate a product within narrow quality parameters and retrieve as much of this product as possible from the input flow.

The choice of separation technologies depends on the physical properties distinguishing the product from the remainder of the flow to be separated. Separation technologies may be a major cost factor, especially for high price products and products from biotechnical conversion technologies.

- *Mechanical separation technologies*

  If the materials to be separated can be distinguished either by their mechanical properties (e.g. particle size and density) or varied physical states (e.g. liquid and solid), then mechanical separation technologies such as sieves, cyclones, centrifuges and presses offer simple and relatively cheap options to separate them. Applications of mechanical separation technologies are seen in retrieving oil from crops (presses) or separating grains from husks (cyclones).

- *Membrane technologies*

  Membrane technologies are gaining importance within bioresource value chains. They separate according to size as well as interaction with surfaces and range from simple filters to ultrafiltration, nanofiltration, reverse osmosis and electro dialysis. The application of membrane technologies ranges from simple separation of solid particles from fluids (filtration and ultrafiltration) to separation of charged compounds (reverse osmosis) and complex molecules (nanofiltration, electro dialysis). An important advantage of membrane technologies is easy scalability since their capacity is directly proportional to the membrane area, making them easily applicable even at smaller scales.

- *Thermodynamic separation technologies*

  If fluid materials are distinguished by their thermo-physical properties (vapour pressure, solubility, surface interaction etc.), these differences can be exploited by technologies used to separate them. There is a wide range of thermo-physical properties that may be used, like vapour pressure (distillation and rectification), solubility of gaseous (absorption) and liquid (liquid–liquid extraction) compounds and surface interaction (adsorption, chromatography). In all cases, a separation agent must be employed such as heat (distillation, rectification), a solvent (extraction, absorption) or an adsorbent (adsorption, chromatography). The applications of thermodynamic separation technologies range from separating low molecular weight products from liquids (distillation, rectification, liquid–liquid extraction) and gases (absorption, adsorption) to the separation of more complex compounds (adsorption, chromatography). Most technologies become more efficient with size, making them ideal for large-scale, central plant applications.

- *Separation technologies for speciality products*

  High price, low volume products require particular care in their separation and polishing. Their higher price allows for more complex and costlier separation methods. For these products, separation costs are often the single most important cost factor in the value chain. Technologies of choice are steam distillation (e.g. for separating essential oils from herbs), supercritical extraction and chromatography. These technologies usually operate in a 'batch' mode and at relatively low capacities.

  Table 5.6 provides an overview of the characteristics of production technologies.

5.2.3.6 Rear Guard Technologies

In order to fully realize the ecological advantage of bioresource use, the residues of products generated by bioresource based value chains have to be re-integrated into the biosphere at the end of their life cycle. The carbon and hydrogen contained in the products may be easily re-integrated into the global carbon and water cycle sustainably. Using the nutrients (e.g. nitrogen, phosphor, calcium, etc.) contained in the original bioresources for maintaining the fertility of the basic resource, i.e. land, will finally allow sustainability of these value chains. The management of these nutrients is therefore decisive for the ecological success of an economy based on the increased use of bioresources. There are technologies available that support the management of organic residues from bioproducts. Their objective is to transform the residues into materials that may easily be returned to the land and conditioned to provide maximum ecological benefit. Table 5.7 provides an overview of the characteristics of these technologies.

- *Combustion of dry organic solid residues*

  The carbon and hydrogen content of residues are oxidized to form CO_2 and water, and as such can be re-integrated into natural global cycles. Mineral compounds are bound in ashes. As already discussed, combustion is always a separation step for heavy metals which are volatile at higher temperatures and accumulate in the small fraction of fly ash. Other contaminants like nitrogen oxides and chlorine have to be retrieved from the flue gas with appropriate end-of-pipe technologies. The fraction of ashes not contaminated with heavy metals can be returned to agricultural land, replenishing micro-nutrients. Contaminated ashes concentrated with heavy metals must be securely disposed of. A major advantage of combustion is the provision of heat that can be utilized. Due to the complexity of the necessary end-of-pipe technologies to avoid ecological pressures, the combustion of organic residues requires medium- to large-scale plants.

Table 5.6 Characteristics of production technologies

Type	Technology	Resources	Products	By-products	Size	Remarks
Extractive	Steam distillation, extraction, pulping	All primary and selected secondary bioresources	Valuable ingredient	Secondary bioresources	Small to industrial	Stringent quality requirements for raw materials
Thermal conversion	Combustion	Dry primary and secondary bioresources	Heat	Ashes (high transport densities, nutrient rich)	Very small to' industrial	Rear guard technology
	Pyrolysis	Dry primary and secondary bioresources	Low-grade gas, pyrolysis oil, char coal	Ashes	Small to industrial	Broad quality margin for raw materials
	Gasification/Fischer–Tropsch synthesis	Dry primary and secondary bioresources	Hydrocarbons	Ashes	Industrial	Gateway technology
Biotechn. conversion	Submerged bioconversion	Bioresource derived substrates	All Organic compounds	Liquid/wet secondary bioresources	Industrial (due to downstream)	Pre-treatment of resources necessary, gateway technology
	Solid state bioconversion	Humid primary and secondary bioresources	Intermediate products	Humid secondary bioresources	Very small to medium	Robust, front line as well as rear guard technology
Chemical and enzymatic conversion	Chemical conversion	Particular ingredients	Simple chemicals	Small amounts	Industrial	Cheap, high conv. efficiency
	Enzymatic conversion	Particular ingredients	Complex chemicals	Small amounts	Industrial	High conv. efficiency
Separation Technologies	Mechanical and thermod. Separation techn.	Various	Desired products	Undesirable parts of input	Small to industrial	Decisive cost factor for high price products

Table 5.7 Characteristics of rear guard technologies

Technology	Effect	Input	Valuable Products	Considerations
Combustion	Oxidizes carbon and hydrogen content	Dry organic residues	Heat, ash	Separation of heavy metals to small fly ash fraction
Composting	Microbial decomposition of residues	Humid solid organic residues	Compost	Quality of compost defined by contaminant free input
Anaerobic digestion	Microbial production of biomethane in submerged bioconversion	Wet or liquid organic residues	Biomethane, digestate	Limits to digestate dispersion on land, high transport effort for digestate logistics

- *Composting*

  Humid organic residues may be treated by composting, which is a form of solid-state bioconversion. The product is a material that improves soil quality and water retention. Care must be taken to ensure that the residues subjected to composting are free of contaminants as there is no way to clean the resulting compost.

- *Anaerobic digestion*

  Anaerobic digestion is a kind of submerged bioconversion capable of treating very humid (e.g. municipal sewage sludge) and liquid (e.g. manure) residues. A considerable part of the carbon content of these residues is transformed into biomethane, with the remainder ending up as CO_2 in the biogas product. Phosphor and nitrogen compounds remain in the liquid effluent referred to as digestate. When digestate is spread on land, these compounds may be returned to agricultural land. However, care must be taken not to overload land by spreading the digestate as some nutrients available in water soluble form may easily migrate to ground water. As the digestate has low concentrations of nutrients, there will be limitations faced in its deployment on agricultural land; therefore, the transport effort for digestate logistics is considerable. The advantage of anaerobic digestion is that it generates biomethane. This is an example of a trivalent energy technology that provides either gas that may be (after cleaning) injected into the natural gas grid or heat and power if the gas is burned in a CHP. Upgrading biogas for the injection, however, is costly.

5.3 Conclusions

Bioresources are very diverse resources in terms of composition, quality and origin. One unifying aspect of bioresources is, however, that they are contextual; their emergence and quality depending on spatial, cultural and the economic framework. Another general aspect of value chains of bioresources is that logistics plays a much more important role than for the fossil resources with which they compete, in many applications.

Due to their often disadvantageous logistical properties like low transport densities, large humidity and degradability, de-centralized technologies in particular for conditioning and storing are often part of value chains based on bioresources. Increased effort to collect, transport and store bioresources may contribute to cost. The wide range of valuable content of many bioresources may give raise to interesting by-products, adding revenue to the value chains.

In many applications, the incredible synthesizing power of nature will generate materials that may only be produced by complex technologies, if non-bioresources are used. Yet, even if the desired compound is not already present in bioresources, there is a wide variety of conversion technologies available that will provide it. As a rule, thermal conversion technologies will render relatively simple platform compounds; chemical and enzymatic conversions will produce valuable chemicals with high yield; while bioconversions have the widest possible range of products, they will convert raw materials with lower yields.

A major advantage of bioresources is that their utilization may close natural material cycles, thus avoiding global environmental threats like climate change. In order to realize this advantage, so-called rear guard technologies have to be employed. These technologies will convert waste flows, from bioresource-based value chains to materials that may safely be integrated into the environment.

Using bioresources requires strict management of material flows as well as a closely interlinked network of de-central and central technologies. As they are contextual resources, their utilization must be adapted to retain or improve the quality of this context, in particular the functioning of productive ecosystems, regional economies and societies.

References

Ecker J, Schaffenberger M, Koschuh W, Mandl M, Boechzelt HG, Schnitzer H, Harasek M, Steinmüller H (2012) Green biorefinery Upper Austria-pilot plant operation. Sep Purif Technol 96:237–247

Hildmann C (1999) Temperaturen in Zönosen als Indikatoren zur Prozeßanalyse und zur Bestimmung des Wirkungsgrades. Energiedissipation und beschleunigte Alterung der Landschaft. Dissertation Technische Universität Berlin, FB Umwelt und Gesellschaft. D 83. Mensch & Buch, Berlin

Kamm B, Hille C, Schönicke P, Dautzenberg G (2010) Green biorefinery demonstration plant in Havelland (Germany). Biofuels Bioprod Biorefin 4(3):253–262

Narodoslawsky M (2016) Towards a sustainable balance of bio-resources use between energy, food and chemical feed-stocks. Found Trends® Ren Energ 1(2):45–107

Zhu X-G, Long St P, Ort DR (2008) What is the maximum efficiency with which photosynthesis can convert solar energy into biomass? Curr Opin Biotechnol 19:153–159

Chapter 6
Bioresources for Distributed Energy Systems

Michael Narodoslawsky

Bioresources require as a basic resource (fertile) land to convert carbon dioxide, water, and solar irradiation into biomass. This makes them inherently de-centralized resources whose productivity is limited by the availability of these valuable basic resources, namely fertile land and water. Both of these basic resources are limited, which makes it crucial to utilize them with the highest possible efficiency. In practice, it means that all parts of bioresources harvested as well as all by-products along the value chain employing bioresources must be utilized, either for obtaining economic and societal profit or for maintaining the fertility of the basic resource (land) by returning nutrients to agricultural land or forests.

6.1 Regionally Adapted Biorefinery Systems

The technological principle emerging from this strict requirement of the utilization of all the biomass taken from nature as input to our economy is the *biorefinery concept,* first introduced by Carlsson (1983). Following the system proposed by the IEA,[1] a biorefinery utilizes multiple feedstocks in a system of conversion processes linked by platforms (intermediate chemicals and energy) to generate a portfolio of products (see Fig. 6.1).

The general idea of biorefineries is to treat and utilize multi-stock *feedstocks* (e.g., crops like wheat and their harvest residues like straw) in technology systems consisting of *conversion technologies* (e.g., combustion, gasification, fermentation, chemical conversion, and various separation processes) that are interlinked by *plat-*

[1] The International Energy Agency (IEA) dedicated its Task 42 to the definition of bio refinery systems. The results of this Task are available at https://subsites.wur.nl/en/ieabiorefinery/Publications.htm. [Dec. 2017]. The current text follows the logic laid down in IEA Task 42.

M. Narodoslawsky (✉)
Institute of Process and Particle Eng, Graz University of Technology, Graz, Austria
e-mail: narodoslawsky@tugraz.at

Y. Krozer and M. Narodoslawsky (eds.), *Economics of Bioresources*,
https://doi.org/10.1007/978-3-030-14618-4_6

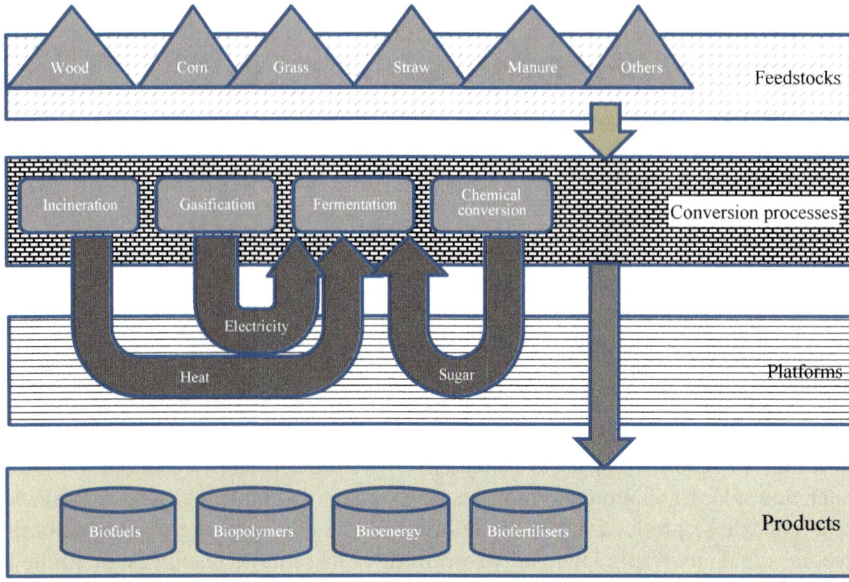

Fig. 6.1 The biorefinery approach to utilising bioresources

forms (intermediate flows like process heat, sugars, organic acids) and generate a broad portfolio of *products*. The overarching goal of biorefineries is to utilize all input flows completely. In order to exemplify the contextual nature of biorefinery systems, we take a closer look into one particular representative, the Green Biorefinery.

In this biorefinery, grass is ensiled and the silage then mechanically separated in a liquid and fibrous phase. The liquid phase runs through a separation chain, extracting lactic acid, and amino acids as products that may then be further refined. The fibrous phase may be conditioned into fiber products (e.g., boards) used for animal feed or as feedstock in a biogas CHP plant together with the remainder from the liquid phase and manure, which is the second feedstock for this biorefinery. At each step, fine high value chemicals may be separated from the platform flows, linking the main conversion technologies. Figure 6.2 shows a number of products.

As bioresources are inherently contextual, dependent on natural endowments on a regional scale, feedstocks may vary from region to region. Utilization of feedstocks in biorefineries depends furthermore on the existing structure of demand, technical infrastructure, and skills of the workforce. Therefore, structures of biorefinery systems are strongly dependent on the regional context and must be adapted to optimally suit the available bioresources, pre-existing conversion technologies, and market demand as well. The Green Biorefinery, for example, fits well into a region that has surplus grassland (providing feedstock for ensiling) and considerable livestock breeding to provide manure. This type of biorefinery is also compatible with biogas CHP plants that exist regionally.

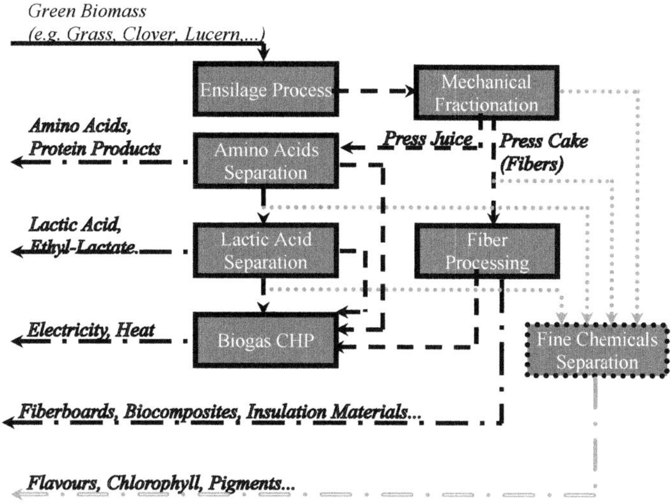

Fig. 6.2 Example for biorefineries: Green Biorefinery

As demonstrated from this example, the concept of biorefineries is just one part of the answer to the challenge of the contextual nature of bioresources. Rather than thinking in terms of linear *production chains* based on particular resources, biorefineries offer interlinked technology systems to cope with the diversity of regionally available bioresources and technological infrastructure. It is important to note at this point, that contrary to fossil resource based refineries, biorefineries are neither single site industrial installations, nor are they standardized technological systems. Parts of these biorefineries can be realized in small-scale de-central and geographically distributed sites, while other parts may be realized in a larger scale at central industrial sites. The geographical and technical structure of biorefinery systems must be adapted to the region's unique natural endowment, technical infrastructure, and socioeconomic setting. This forms the *bioresource fingerprint* of a region shaped by the available bioresources, existing technical infrastructure, local demand (especially for heat), and skills of the regional workforce.

6.2 Logistics for Optimizing Scale and Distribution

A major factor in shaping this fingerprint is the logistics of bioresource collection and product distribution, as well as capacity and time dependency of both resources and product markets. Bioresource collection becomes crucial for lower grade primary bioresources (e.g., grass) and secondary bioresources from agriculture (harvest residues, manure, etc.) as these feedstocks usually have high humidity and/or low transport densities (see Chap. 3 for typical data). However, these feedstocks are pre-

Table 6.1 Energy density and transport energy for selected bioresources

	Energy density (MJ/m^3)	Transport energy (kJ/MJ km)			
		Tractor[a]	Lorry[b]	Train (electric power)[c]	Ship (river)[b]
Light fuel oil	35.800	0.031	0.03	0.004	0.004
Wood pellets	10.800	0.075	0.08	0.011	0.012
Corn	10.000	0.079	0.09	0.012	0.012
Wood logs	7.000	0.097	0.09	0.012	0.015
Wood chips	3.100	0.149	0.10	0.015	0.025
Straw bales	1.990	0.232	0.13	0.021	0.039
Straw loose	970	0.425	0.19	0.034	0.072
Manure	700[d]	1.89	1.9	–	–
Critical density (kg/m^3)[e]	–	720	300	390	920

[a]Tractor data: http://deutzfahr.at/fileadmin/Bilder/Prospekte/Agrofarm_TTV_Profiline.pdf. [January 2017]. Diesel consumption for 75 PS tractors from: Gastinger G.: Untersuchung des Kraftstoffverbrauchs in der 75 kW Traktorenklasse mit einem leistungsverzweigten und lastschaltbaren Getriebe, Master Thesis, University of Life Sciences Vienna, 2011 available from http://epub.boku.ac.at/obvbokhs/content/titleinfo/1127129. [January 2017].Data for trailer from: http://www.hb-brantner.at/fileadmin/user_upload/fahrzeugbau/files/Preisliste/PL01-2013-AT-web.pdf. [January 2017]

[b]Borken J., Patyk A., Reinhardt G.A.: Basisdaten für ökologische Bilanzierungen: Einsatz von Nutzfahrzeugen in Transport, Landwirtschaft und Bergbau, Vieweg & Teuber, Wiedbaden, 1999

[c]Data from: http://www.forschungsinformationssystem.de/servlet/is/342234/. [January 2017]

[d]Calculated as energy from methanization

[e]Calculated as max. load weight divided by max. load volume

destined for technical utilization and energy provision as they are widely available, and there is no demand for these materials from the food sector.

Actually, tractors are the most common *first mile* means of transportation for secondary bioresources from agriculture, like harvest residues and manure. This is a limiting factor for the distance these feedstocks may sensibly travel during this first leg of collection. Tractors are particularly inefficient as they show a low ratio of payload to overall weight. Moreover, tractor trailers have limited volumes, resulting in a limitation of transport volume for densities below 720 kg/m^3. Table 6.1 shows the required transport energy per MJ energy content for various means of transportation of selected bioresources. The numbers in this table take the volume limitation for materials with low densities into account.

As can be calculated from Table 6.1, manure may be transported with a tractor for 5.3 km and loose straw for 23.5 km using about 1% of the energy contained in the transported feedstock. Such feedstock require de-central treatment, either by utilizing them directly or conditioning them for further transportation, e.g., by compacting, pelletizing, or reloading them on transport means with greater efficiency.

The values in Table 6.1 may be used for a first estimate of the energy required to collect and deliver various bioresources. If collection (per energy unit in feedstock) requires more transport energy, then geographically distributed and small-scale installations are more sensible.

Besides collection of feedstock, logistical considerations of distributing products are also crucial factors in shaping the size and geographical distribution of regionally adapted biorefinery elements. Many conversion technologies generate by-products that contain valuable nutrients that must be brought back to agricultural land in order to maintain their fertility. This, for example, applies to biogas CHP which generates biogas manure, or biomass combustion and gasification generating ash. Especially for aqueous by-products such as biogas manure, transport energy is considerable and short distribution distances from conversion sites to agricultural land are favorable.

To choose the right size and site for bioresource conversion technologies, it is crucial to estimate the transport requirements for the collection of bioresources and—if applicable—distribution of nutrient containing by-products (e.g., biogas manure or ash) to be brought back to agricultural land. The following formulas may guide an initial rough estimate. Whether the calculations shall be carried out on the basis of mass or volume would depend on the transport property of the resource/by-product, and the critical density of the means of transport (see Table 6.1). If the transport density is larger than the critical density, calculations should be based on mass, otherwise on volume.

Let C_c be the feedstock (or by-product, if applicable) capacity of a certain conversion technology in t/y or m^3/y (whatever dimension is more appropriate). Let y (t/ha.y or m^3/ha.y) be the yield of the bioresource or the allowed distribution rate of a nutrient by-product; f_l the fraction of land on which this resource grows in the region or on which the by-product may be distributed. If the whole harvest in the region is not at one's disposal for the conversion plant or if all suitable land is not available for spreading the nutrient by-product, then f_a represents the fraction of the bioresource that actually serves the plant.

$$A_c = C_c/(y.f_l.f_a) \tag{6.1}$$

In Eq. (6.1) A_c then designates the total regional area that provides for the plant or is served by the by-product. A rough estimate of the mean transport distance d_m can then be made by:

$$d_m = \frac{2}{3}\sqrt{\frac{A_c}{100 \cdot \pi}} \tag{6.2}$$

Bear in mind that Eq. (6.2) is just a rough estimate that does not take into account topographical considerations in the real region!

With the capacity of the means of transportation, C_t the number of single runs, N_r to supply the plant or distribute the by-product can be calculated:

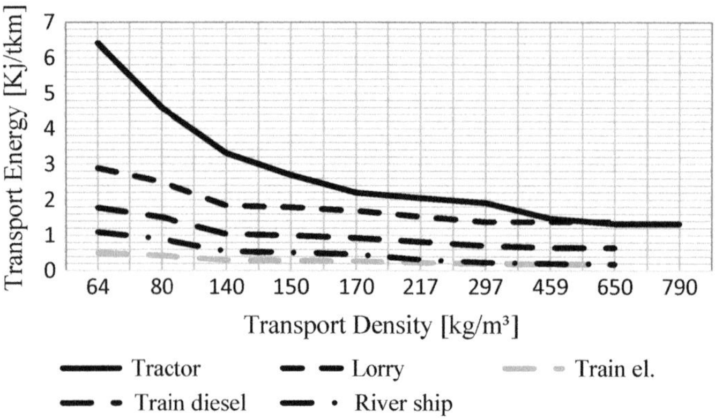

Graph 6.1 Transport energy per tkm versus transport density for different means of transport

$$N_{\mathrm{r}} = \frac{C_{\mathrm{c}}}{C_{\mathrm{t}}} \tag{6.3}$$

The distance of single runs per year can then be calculated as follows:

$$d_{\mathrm{t}} = N_{\mathrm{r}} \times d_{\mathrm{m}} \tag{6.4}$$

Graph 6.1 shows energy values of transporting bioresources.

Using values from Graph 6.1, the transport energy per year can be roughly estimated with d_{t}. Bear in mind that in many cases, this distance has to be estimated as double as the (often empty vehicle on the) return journey must be factored in as well. As a rule of thumb, the lower the quality of feedstock and larger the distribution effort (energetic and infrastructure) for by-products, means the conversion technologies should become smaller in size and more distributed.

Particular care is required to integrate elements of biorefineries into regional energy systems. Conversion technologies within biorefinery systems can provide a wide range of energy services and energy carriers in various combinations. Small combustion units may provide just enough heat of medium temperatures, whereas larger units may be operated as CHP systems with an increasing share of power the larger their capacity is. Digestion may generate biogas that can be used as feedstock for CHP plants but can also be cleaned and distributed via the natural gas grid. Gasification may generate feedstock for CHP units but also be operated in a way to produce synthesis gas that can be converted into synthetic natural gas (SNC) or to biofuel via Fischer–Tropsch synthesis (FT-biofuel). The technologies chosen in a regionally adapted biorefinery system depend on the energy demand structure in that region, temporal load profiles, and the existing energy distribution infrastructure.

Different energy forms vary widely in their respective logistical properties. This applies to distribution losses as well as to their feasibility and cost of storage. Liquid

Table 6.2 Transport losses for energy distribution grids

Energy form	Means of transport	Loss per 100 km (% of energy transported)
Crude Oil	Trunk pipe line	0.036[a]
Natural gas/biomethane	Trunk pipe line	0.15[b]
Electricity	10 kV grid	30
	380 kV (free standing)	1.1[c]
	380 kV DC	0.3[d]

[a]According to Konstantin (2009)
[b]According to Kennedy (1993)
[c]According to Oswald (2007)
[d]According to http://www.weltderphysik.de/gebiete/technik/energie/speichern-und-transportieren/strom/hochspannung/. [January 2018]

Graph 6.2 Heat loss versus annual heat demand per m distribution pipe for high and low temperature systems

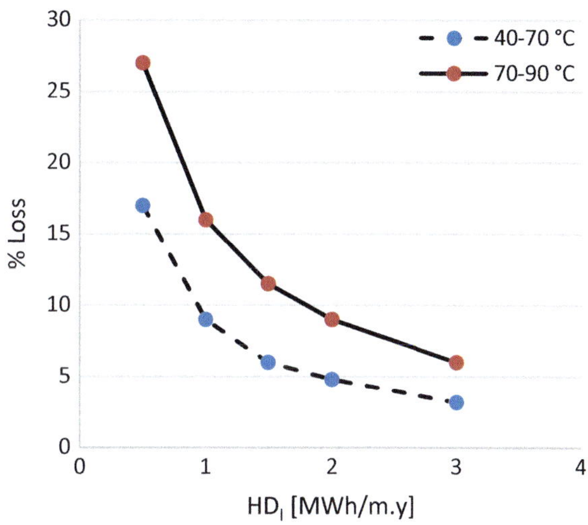

energy carriers like biofuel, gaseous products like biomethane and electricity may be transported over long distances with relative ease (see Table 6.2).

In contrast, heat is an energy form that may only be used close to its emergence. Losses in heat depend on its annual transport through the grid, and the temperature level of the heat.[2] Graph 6.2 shows this dependency. The linear heat density $\mathrm{HD_l}$ in this graph is defined as

$$\mathrm{HD}_l = \frac{\sum h_\mathrm{i}}{l} \qquad (6.5)$$

[2]A thorough analysis of bioenergy-based heat distribution systems can be found in IEA bioenergy task 32. Nussbaumer and Thalmann (2014) provide an overview on existing bioheat distribution systems. All available from http://www.ieabioenergytask32.com/. [January 2018]

The annual heat demand of customer I is h_i, and the length of the distribution pipe is 1.

Graph 6.2 shows that lower temperature levels of distributed heat results in lower transport losses. It also very clearly indicates that short distribution distances and large annual heat demand are crucial for the efficiency of bioenergy-based heating systems.

The disadvantage of transport losses regarding heat from bioresources is somewhat compensated by the low cost of storing heat compared to electricity. This allows bioresource-based CHP systems to operate following the demand of the electricity grid: whenever electricity demand exceeds production, bioresource CHP systems may operate to fill the gap. The heat produced as a coupling product is stored and provided to customers according to heat load profiles that may differ widely from electricity demand profiles.

This integration into distribution grids may even be carried further with conversion technologies that can operate as CHP systems and alternatively biomethane providers, such as digester plants (for wet feedstocks) and gasification plants coupled with SNG conversion units (for dry feedstocks). These trivalent plants can serve the gas distribution grid as long as no electricity or heat is demanded and may operate as CHP systems to fulfill the demands of the electricity grid and store heat otherwise. This ability to negotiate between different energy distribution systems makes bioenergy units particularly important elements of regionally adapted biorefineries. This importance will only grow when the whole energy system becomes more dependent on renewable energy sources like wind power and photovoltaic, as these sources show strong time dependency in their production. Bioenergy technologies, if optimally integrated, can provide invaluable services in stabilizing distribution grids while efficiently utilizing available bioresources.

Summarizing these arguments, integrating biorefinery elements (and in particular conversion processes) that can optimally provide various forms of bioenergy to a region is a complex planning task. The following guidelines can help to structure regional bioresource utilization systems (Narodoslawsky 2016):

- *'Refinerize' existing bioresource utilization sectors*
  This allows us to draw on existing logistical systems to collect bioresources, as well as on the existing regional know-how and workforce while broadening the product portfolio based on regional bioresources. In this manner, we can utilize resources more efficiently.
- *Give priority to material goods*
  Bioresources are the only way to convert solar radiation into material goods. This will, in the long run, require a bio-based economy wherein goods of daily consumption (or with a short time of use) will be predominantly produced from bioresources. Besides this long-term objective, material goods usually allow for longer, more complex value chains which in turn provide better income chances for regional economies.

- *Retain as much material close to productive land*
  Preserving productivity of agricultural land and forests is a major concern for optimally utilizing bioresources. It is therefore necessary to return nutrient by-products to the land if they are not otherwise used in the value chain. As most of these by-products have disadvantageous logistic properties, transport distances between the point of their emergence and the land to which they will be returned must be kept short. This increases the necessity for de-central and distributed elements of regional biorefineries, in particular energy providing units that are based on low-quality secondary bioresources such as harvest residues and agricultural waste products.
- *Plan bioenergy units at intersections of distribution grids*
  A major advantage of energy providing elements of biorefineries is their ability to negotiate between different distribution grids; thus optimizing their service as stabilizing elements for a sustainable energy system in the future. In order to minimize transportation losses, these units must be located at the intersections of distribution grids.
- *Optimal local heat integration is the key to efficient bioenergy systems*
  Heat is the one energy service from bioenergy plants that must not be transmitted over long distances. Utilizing as much heat as possible, however, is paramount to efficient resource use and economic viability as well. In particular, this requires continuous or at least long-term consumers of heat close to plants that provide bioheat. Examples for such especially interesting customers include industrial processes operating all year round, food processing, and drying plants for bioresources as well as crops. Greenhouses may also be interesting customers too. The lower the temperature level of the heating system of the customer the longer may be the transmission line as low-temperature distribution systems have considerably less energy losses (see Graph 6.2).

6.3 Planning of De-central and Distributed Systems

Bioresources are inherently contextual, so planning their utilization requires taking into account the context of regional climate and primary production; local and regional demand (in particular of energy); regional industrial structure, expertise, and workforce (Narodoslawsky 2017).

Fossil resources are commonly converted and utilized in big, centralized units following the principle of 'economy of scale'. This means that larger fossil resource-based units are inherently more efficient than smaller ones. The reason for this is that fossil resources usually have excellent logistical properties, like high transport density and low or even zero humidity. They also emerge from large point resources like coalmines, oil, and gas fields. This reduces the impact of logistics on the structure of fossil-based industries, leading to the predominance of technological efficiency (and thus size) as the guiding principle for planning.

As already discussed above, logistical considerations play a major role in determining the site and size of the elements of regional bioresource utilization systems. The appropriate planning tools are required to adapt bioresource-based energy plants optimally to their regional feedstock potential; plan existing industrial and energy distribution infrastructure as well as local and regional energy demand that takes into account the transport of feedstock and by-products. The crucial challenge is finding the optimal structure of a technology network that maximizes regional revenue from bioresources while minimizing ecological impact (Narodoslawsky et al. 2016). This in turn requires powerful structural optimization methods. One method that has proved itself particularly useful for this task is Process Network Synthesis (PNS) (Cabezas et al. 2015), based on the P-graph method proposed by Friedler et al. (1995). A full explanation of its mathematical approach to structural optimization is out of the scope of this chapter. A general introduction to the use of PNS as well as its mathematical background can be accessed on the PNS homepage.[3] This is a general structural optimization method developed for process industries that use combinatorial rules to generate a maximum structure encompassing all possible ways to provide defined products and services, from a given set of feedstocks using a defined set of technologies.

For the purpose of optimizing the utilization of bioresources in a particular regional context, this method may easily be adapted. The feedstocks are all regionally available bioresources, including secondary and tertiary bioresources. For each bioresource, its unit cost and available amount has to be known. Technologies include conversion technologies that either convert these feedstocks into platforms or desired products and services, or those that convert intermediate platforms into products. In contrast to their original application in process industries, technologies may as well be logistical units whose role in the technology network is to provide a given material (either a feedstock or an intermediate platform) at a location different from its point of emergence. For all these technologies, fixed and variable costs as well as capacities have to be known. Similar technologies of varying sizes are, in this method, considered different entries as their cost characteristics and capacities differ.

Products in the application of PNS include all kinds of bioproducts that may be generated from the available bioresources with the use of the technologies considered. They include energy services like heat and electricity, generated either by the direct conversion of a certain bioresource or as a by-product of processes in a value chain leading to a certain bioproduct. For each product, the price at which it can be bought on the market must be known. Using this method, it is possible to assess the limits of product flows (for instance, when local markets for certain products are limited), or to set a certain amount of the product that must be provided (for instance, when the heat demand in a region must be provided from renewable resources).

As mentioned, from the sets of resources, technologies, and products, PNS generates a structure that includes all viable links between resources, technologies, platforms, and products. This all-encompassing structure is the so-called maximum structure. Using a branch-and-bound optimization routine, viable technology networks

[3]Thorough information on the method is available from http://p-graph.com/. [January 2018].

are found as sub-structures of this all-encompassing maximum structure. For these viable structures, costs are calculated and compared, finally leading to the most optimal technology network based on available bioresources that fulfill all energy and material balances as well as all predefined boundary conditions (e.g., local demands that have to be met).

In an extension, this method can also deal with the time-dependent provision of feedstocks and demand for products and services (Heckl et al. 2015). An overview of different case studies using this method for optimizing regional bioresource utilization and integration of bioenergy in regional energy systems is provided by Narodoslawsky (2017). A rough estimate of optimal technology networks adapted to the regional feedstocks and energy demand can be achieved using the RegiOpt software[4] (Kettl et al. 2011) based on the PNS method.

6.4 Conclusions

Low transport density, humidity, and degradability of bioresources imply complex logistics. Bioresource-based value chains have, in particular, a 'first and last mile' problem: this means that the collection of bioresources and first conditioning steps as well as re-introduction of waste flows from value chains often require de-central technologies. These technologies and the logistics they require have to be systemically adapted to their regional context, in order to achieve maximum ecological, economical, and social profit from these resources.

The key to adapting these technology systems to their regions is to integrate energy provision and use: many technologies that condition bioresources for further utilization and transport require energy; while many of the rear guard technologies that treat waste and by-product flows from bioresource value chains provide energy.

Energy comes in the form of heat, electrical power, and carriers of gaseous, liquid, and solid energy. Each of these forms has different distribution systems. The interaction of these distribution systems has a strong influence on the shape of distributed energy systems.

Of the energy forms mentioned, heat can only be distributed over short distances. This means that technologies providing heat and those consuming heat must be placed closely together.

Many technologies based on bioresources generate energy in different forms. Examples are biodigesters that generate biomethane which can be consumed in an internal combustion engine to provide heat and power, or cleaned and injected into the gas grid. This ability to negotiate between different distribution grids makes such technologies particularly important for energy systems that rely on large shares of intermittent renewable energy sources—like wind or solar radiation. Such technologies should, therefore, be located at the intersection of various distribution grids.

[4]The software is available from http://regiopt.tugraz.at/index.php/en. [January 2018]. A general PNS software is available from http://pgraph.dcs.uni-pannon.hu/template/. [January 2018].

Distributed energy systems based on bioresources shall always cover heat, demanded locally. They can, in many cases, export energy in different forms and contribute to attaining stability of distribution grids. As the dependency of the overall energy systems on renewable sources grows, the importance of the distributed, bioresource-based energy systems increases as well.

References

Cabezas H, Heckl I, Bertok B, Friedler F (2015) Use the P-graph framework to design supply chains for sustainability. Chem Engn Progr 111(1):41–47

Carlsson R (1983) Leaf protein concentrate from plant sources in temperate climates. Leaf Protein Concentrates 52–80

Friedler F, Varga JB, Fan LT (1995) Decision-mapping for design and synthesis of chemical processes: applications to reactor-network synthesis AIChE symposium. Chem Eng Sci 50(11):1755–1768

Heckl I, Halász L, Szlama A, Cabezas H, Friedler F (2015) Process synthesis involving multi-period operations by the P-graph framework. Comput Chem Eng 83:157–164

Kennedy JL (1993) Oil and gas pipeline fundamentals. PennWell Publishing Comp, Tulsa

Kettl KH, Niemetz N, Sandor N, Eder M, Heckl I, Narodoslawsky M (2011) Regional optimizer (RegiOpt)–sustainable energy technology network solutions for regions. Comput Aided Chem Eng 29:1959–1963

Konstantin P (2009) Praxisbuch Energiewirtschaft Energieumwandlung,-transport und-beschaffung im liberalisierten Markt. Springer, Heidelberg

Narodoslawsky M (2016) Towards a sustainable balance of bio-resources use between energy, food and chemical feedstocks. Found Trends® Renew Energy 1(2):45–107

Narodoslawsky M (2017) Bioenergy provision: utilizing contextual resources. Current Opin Chem Eng 17:93–97

Nussbaumer T, Thalmann S (2014) tatus report on district heating systems in IEA countries. Prepared for the International Energy Agency IEA Bioenergy Task 32 and Swiss Federal Office of Energy, and Verenum, Zürich. Available from www.ieabioenergytask32.com. [January 2018]

Narodoslawsky M, Cabezas H, St Maier, Heckl I (2016) Using regional resources sustainably and efficiently. CEP 112(10):48

Oswald BR (2007) Verlustenergieabschätzung für das 380-kV-Leitungsbauvorhaben Wahle—Mecklar in der Ausführung als Freileitung oder Drehstromkabelsystem, University Hannover, Hannover. Accessible from www.netzausbau-niedersachsen.de/…/verlustvergleichwahlemecklarfinalv2.pdf. [January 2018]

Part II
Cases in Bioenergy

This part presents four cases of biofuel production. These cases illustrate the possibilities for satisfying rural energy demands and the use of organic residues for energy consumption while maintaining the availability of natural resources due to the better utilization of regional bioresources. First, a local value chain of biofuel is presented to point out cost-effective applications for local heating. Then, people's opinions about the use of bioresources for heating are presented with the objective of demonstrating that tangible local benefits are a key condition for participation. Third, bioresources are linked to the synthetic gas based on renewable energy for heating to underpin the abundance of resources on the national scale. Finally, cross-sectoral linkages between the organic residues are illustrated to show that the circular economy of bioresource is feasible, albeit challenging.

Chapter 7
Biofuels from Willows

Sandor Bartha

The Green Energy Cluster, an organization in Romania with 75 members from business, academia, and the public sector, acts as a business and market developer, an educator and a trainer for specialists, as well as an executor of research, innovation, and facilitator of international cooperation in renewable energy. Its focus is on the use of bioresources for energy in rural areas. Biofuel based on short rotation cultivation (SRC), herewith, provides an economic bioresource which substitutes mineral fuels in a sustainable manner; if marginal land is used and low-input agroforestry is applied. This case shows practices in the Covasna region. Subsequent steps in short rotation cultivation are introduced followed by processing into biofuel and its use in communities, as well as an economic regional strategy based on the experiences.

7.1 Short Rotation Cultivation (SRC)

Romania has a forest reserve of 6,494,728 ha of which 16,520 ha a year (equivalent to 16.78 million m^3 bioresources) is harvested (Baste the statistical dates presented in: Aebiom (2015) and FAO (2010). Forest in the Carpathian area (the Covasna region being part of it) covers about 19% of the whole. It generates about 5.6 m^3 per hectare per year, equivalent of 2.2 dry tons per hectare across all tree species (Popescu 2009)—the 4th highest performance in Europe after Ireland, Austria, and Denmark.

Much higher output per hectare can be obtained when fast growing species are cultivated in short rotation. The short rotation cultivations use mainly willows (*Salix*), Poplar (*Populus*), Eucalyptus (*Eucaliptus*), and Acacia (*Robinia*). There are about 150 plantations of the fast growing species on 1200 ha in the Carpathian, out of which 136 ha is in Covasna County which is the subject of this case. These fast growing

S. Bartha (✉)
BIO-C Green Energy Association, 4 Presei Street, 520064 Sf. Gheorghe, Romania
e-mail: sbarthacv@yahoo.ro

Y. Krozer and M. Narodoslawsky (eds.), *Economics of Bioresources*,
https://doi.org/10.1007/978-3-030-14618-4_7

species, harvested a few years after the start of cultivation, generate a high yield of bioresources within a short period of time (Bartha et al. 2017). This is a major advantage for low-income regions where people cannot afford a long waiting time for generating benefits.

The bioresources from forests are used in the production of round wood, structural timber, laminated panels, lumber, and other conventional wood products; while waste streams are transformed into briquettes and pellets for the biofuel used in heating and the sale of electricity in the national electricity market. Those uses can be sustained due to afforestation, and thanks to the SRC on plantations that aim at biofuels. The producers of bioresources are usually small-scale farmers that possess less than 50 ha land. They produce the bioresources for biofuel as well as for various other uses, particularly for foods which complement local agriculture.

SRC plantations are often integrated with agroforestry. It is an alternative to agricultural monocultures and an opportunity for marginal lands near villages. The marginal lands, herewith, are defined as having poor soil quality for agricultural uses, being unsuitable for housing and suchlike uses, or comprise areas prone to flood (Heimlich 1989; Hart 2001). Romanian regulations allow SRC only on marginal lands based on good preparation, fertilization, and clearing from pests. The use of marginal lands for SRC is an option which contributes to sustainable regional development because it is economic for farmers, enhances carbon sequestration, fosters soil and water quality, and supports ecosystem services if sound soil management is applied (Lal 2004).

Bioresources are supplied to the local energy producer. These can be owners of co-generation installations for heat and power (CHP) which are usually large units. They can also be used for district heating on a small scale, for instance by an individual customer as done so in the Covasna region by a greenhouse in the winter, and a bakery that consumes wood chips in a high efficient boiler. Bioresources are generally wood chips for heating because they are cheap and easy to handle, as compared to CHP for heating and electricity. Although co-generators are usually more energy efficient than district heating boilers because they lose less heat CHP's involve complex operations and grid design; while district heating can be downscaled to rural needs at low costs. Residues from agriculture can be used for biogas production, but such technologies are rarely available in Romania. This implies that about 12.6 million tons biomass per year in Romania is not used cost-effectively, instead left on the land or burned (Scarlat et al. 2011).

7.2 The Production Chain

The production chain of bioresources from SRC covers cultivation, harvesting with storage, processing, and use as biofuel. While these steps are briefly covered, we focus on the cultivation of willows because the Green Energy Cluster gained experience with this particular plant.

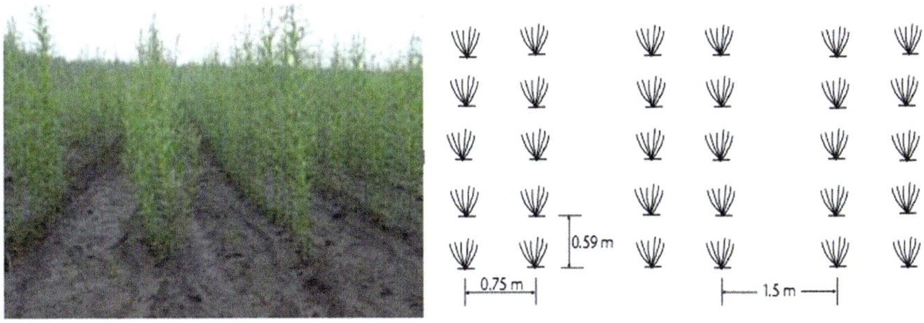

Fig. 7.1 Planting of willows *Salix*, variety *Inger*; photo made by the author

7.2.1 Cultivation

The marginal soil for willows can vary from clay to sand. Willows prefer growing in soil with good aeration, retention of moisture and a pH range of 5.6–7.5. Rainfall of 800–1000 mm per year is sufficient, but retention is important because willows need more water for their fast growth as compared to many agricultural crops. The soil remediation and preparation for cultivation is the first step on marginal land. It covers mechanical pre-plowing, application of roughly 6 liter herbicides per hectare for disinfection, as well as the enrichment of soil with about 450 kg NPK fertilizers per hectare; this amount is reduced to 150 kg NPK in the 2nd and 3rd year. After the harvest each year, additional 20–22 tons sludge from the wastewater treatment plant is used which is controlled for pollution by the local environmental agency.

After fertilization, plants are selected and 14,000 to 15,000 plants per hectare are planted (commercially recommended willow varieties are Tora, Inger, Tordis, Gudrun, Doris, Jorr, and Olof from Sweden). Planting is done in the springtime with a tractor of 150 horse power, while planting designs and densities comprise a double row system developed in Sweden. As shown in Fig. 7.1 plants are set with 1.5–1.8 m between each plant in double rows, 0.75 m between individual rows and 0.55–0.59 m between plants along each row. The first planting in the Covasna region was done manually, but nowadays modern equipment is used.

During the first year, willow grows 3–3.5 cm per day in 1–3 stems that reach the height of 2–3 m. After the first year, the willow plants are cut. That cutting enhances multiple shoots, often 8–10 per plant. After the cutting, coppices enter into a cycle of 2–3 years. In the second year, the shoots grow 6–7 m high and 3–4 cm diameter. Compared to annual crops in agriculture, this cultivation causes less soil erosion and water pollution because of mechanical distortion and lower use of fertilizers and pesticides, respectively; further, more biodiversity and carbon storage is generated.

7.2.2 Harvesting

The harvesting period is from December to middle March. Three methods are used, each with their own pros and cons: direct chip harvesting, whole rod harvesting, or billet harvesting. The direct chip harvesting method enables the production of woodchips on the field itself, followed by storage and drying on aerated platforms near the plantations or it is transported directly to the thermal plant. If the humidity of the wood chips is below 30–40%, they can be used; but there exists a risk of degradation by fungi and bacteria and consequently a loss in calorific value. Whole rod harvesting is used for 2- to 4-year-old plantations, but requires special equipment as it is difficult to handle plants that are 6–8 m long. The billet harvesting method is used for younger plants below 10 mm in diameter and is similar to the harvesting of sugarcane. This is used by the Green Energy Cluster.

Willow can be coppiced eight to ten times, which implies a 20- to 22-year life span of plantations in a short cycle of 2 years. The life span of a plantation can be made longer when 4- to 5-year harvesting cycles are applied; though agricultural and meteorological conditions must be suitable (technical parameters of the harvested biomass are shown in the SR EN ISO 17225-4). Based on the guides for short rotation cultivation of willows (Toivonen and Tahavanainen 1998; Teagasc, AFBI 2010) and experiences of the Green Energy Cluster, the life span of plantation involves a few cyclic activities: pre-plowing once, plowing twice, disk once, planting once, rolling once, harvesting 7 times, and the application of nutrients and herbicide 8 times.

The average yield of willows varies across 58 European regions, ranging between 15 and 25 dry tons per hectare (Nonhebel 1997). Yields in the Covasna region are 15–20 dry tons per ha per year, equivalent of 35 tons per hectare with 45–55% moisture which is typical of the harvest. Following this, the harvested wood is naturally dried to 18–20% moisture. Although 8% moisture is optimal for combustion, wood is not further dried because this requires energy. Wood chips could be dried in large storage pills to avoid degradation but this needs much energy. Wood chips and wood dust could be dried and compressed to briquettes and pellets for optimal combustion, but these processes are costly.

The calorific value of biomass indicates how much energy can be gained from harvested bioresources. Table 7.1 shows the calorific value for energy production, of the main types of biomass compared with fossil fuels.

Fossil fuels have a higher energy density than biofuels but are exhaustible when compared to renewable bioresources. The calorific value of bioresources is determined by humidity. The calorific value of freshly collected willow with 45–55% moisture is 6–8 MJ per kg; when dried to 10–20% is 14–16 MJ per kg and that of dry wood chips are 19 MJ per kg but the latter is only theoretical. Figure 7.2 shows the relation between humidity and the calorific value of plants.

The combustion of bioresources from willow also generates 0.9–2% ash, which is low when compared to coal. Nonetheless, the environmental impacts of the former are relevant because difficult for reuse on fields. When this ash dissolves in water, it

Table 7.1 Typical energy outputs per ton feedstock

	Water content %	Calorific value (GJ)	Energy content (kWh)	Heating oil equivalent (kg)
Hardwood, Beech (naturally dried)	35	11.1	3085	259
Softwood, Spruce(naturally dried)	35	11.3	3139	264
Wood Pellets (kiln dried)	<10	17	4725	396
Wheat Straw (naturally dried)	15	14.4	4032	339
Miscanthus (naturally dried)	15	14.9	4004	336
Energy Willow	30	12.2	–	–
Fuel Oil	Negligible	42.7	11,860	1000
Coal	Negligible	26.7	7417	625

has a high pH value of 10.5 when compared to a pH of 5.5–7.5 for the cultivation of willow.

The economy of SRC with willow depends on the scale of operations because the scale reduces the cost and sales price of the bioresources. Given that many farmers in the Covasna region possess smallholdings of 0.5–15 ha lands, farmers cooperate; for instance, on average one tractor is used on 33 farms (INS, Anuarele Statistice ale Romaniei). Cooperation in SRC is also needed because the minimum area is 2–5 ha per plantation. Smaller and irregular fields are difficult to manage, involve laborious harvesting and costly logistics as only small batches can be transported. Another example being that the usual harvesting machines can cover 5–6 ha per day and are hired by farmers from specialized firms.

The establishment of a short rotation cultivation of willows costs between € 1700 and € 2000 per hectare and throughout its life span delivers 20–25 tons of dry wood at a sales price of € 50– € 70 per ton of bioresource. It implies a payback of 5–7 years, taking into consideration the additional costs of operation for cultivation and harvesting as well as the transport and drying of the wood. It is a reasonably profitable investment. The sales depend on the categories of moisture whose percentages are expressed in alpha numerals such as M20, M30, or M50. The particle size is also important, for instance, the alpha numerals P16, P31.5, P45, P63, and P100 express the diameter of woodchips in mm (Hart 2001).[1] For example, the sales price of wood chips with 30% moisture and 45 mm particle size is about € 58 per ton. Further,

[1]The diameter of particles is standardized in six classes: > 100 mm, 100–63 mm, 63–45 mm, 45–16 mm, 16–3 mm, < 3 mm (CEN/TS 14961).

when we include the national support of bioenergy production, income amounting to about € 70 per ton can be generated. Wood chips with greater moisture content and particle size have a lower price.

7.3 Regional Use

The regional demand for district heating based on bioresources is analyzed. It focuses on offices, schools, small hospitals, and such buildings meant for communal activities, because these are often centrally located. This central location enables a distribution of heat to other houses entailing a district heating network based on bioresources. Demands for heating are large because a majority of such buildings in rural areas were constructed 40–50 years ago (some even 80 years ago) and more so without insulation. Romanian insulation—even today—remains in the initial stages of development.

7.3.1 Energy Performance

In Romania, the energy performance of communal buildings is standardized according to climatic zones and measured as energy demand per m^2. Given the low energy performance in rural areas of the Covasna region, the demand for heat, S_q, is high and varies from 150 to 350 kWh per m^2 a year. Having long winter periods with temperatures as low as −25 °C, the region is located in zone V based the Romanian normative for the thermo technical calculation of buildings, (HG 386/2016) where the demands to be fulfilled are 1600–1800 h a year. Table 7.2 shows energy demands per type of communal building based on the standards.

It is sufficient data to estimate the required boiler capacity. See Box 7.1.

Table 7.2 Heating demand of buildings in the Covasna region in S_Q meaning space heat demand in kWh/m^2 per year

Climatic zone	Period to implement	Collective building	Office building	Building for education	Building for hospitals
IV −21 °C	2015	152	107	192	190
	31.12.2018	144	89	172	149
V −24 °C	2015	178	127	210	214
	31.12.2018	152	98	192	174

Box 7.1 Method for estimation of the required boiler capacity The boiler capacity P_T in kW is based on the building area A, and heating needs S_q during 1500–1600 h a year.

$$P_T = \frac{A \cdot S_Q}{t}$$
$$[\mathrm{kW}]]$$

The yearly heat demand of a building is

$$Q_T = t \cdot P_T$$
$$\left[\mathrm{kWh/year}\right]$$

After correcting for equipment and system efficiency

$$Q_R = \frac{Q_T}{1000 \cdot \eta} = \frac{t \cdot P_t}{1000 \cdot \eta}$$
$$\left[\mathrm{MWh/year}\right]$$

where η is the biomass conversion system efficiency assumed above 90% for the high performance boiler and 85% for district heating.

The humidity of biofuel is essential because moisture content reduces heating. Figure 7.2 shows the relation between the calorific value of woodchips and humidity. It shows a linear decrease in the heating value as humidity increased.

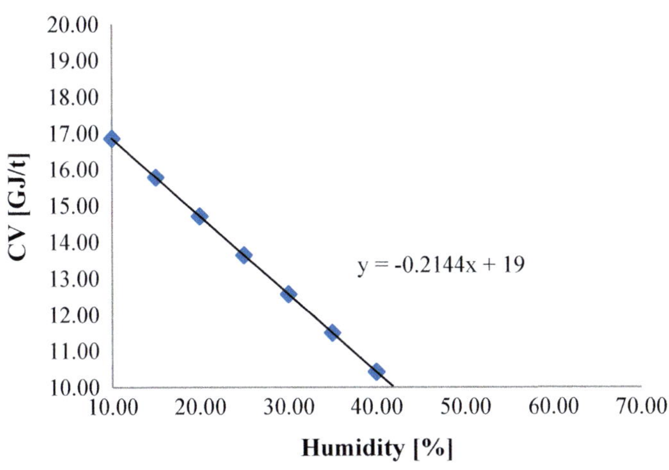

Fig. 7.2 The relation between the calorific value of woodchips and humidity

Table 7.3 District heating system based on biofuels in Covasna: a scenario

Parameter of the bioenergy heating system, operation time 1600 h	Public building	Office building	Education building	Hospital building	Yearly cumulative values
Building useful heated area (m^2)	300	200	500	250	1250
Heat demand ($kWh/m^2/year$) in buildings with insulation	178	107	192	190	
Total useful heat demand (MWh/year)	53.40	21.40	96.00	47.50	218.30
Total useful heat demand (GJ)	192.24	77.04	345.60	171.00	785.88
Biomass moisture content (%)	30.00	30.00	30.00	30.00	–
Biomass calorific value (13–16 GJ/t)	13.00	13.00	13.00	13.00	–
Biomass Need [t/year]	14.79	5.93	26.58	13.15	60.45
Biomass Yield of SCP in t/ha	20.00	20.00	20.00	20.00	–
Marginal Land for SCR in ha	0.74	0.30	1.33	0.66	3.02
Capacity Heaters [kW]	33.37	13.38	60	29.69	136.44

Table 7.3 presents results of the analyses as per the type of building in the Covasna region. The assumptions for calculation of the heating requirements are 1600 h per year operating on full capacity of boilers and the use of wood chips from willow with moisture below 30%.

A boiler capacity of 137 kW is needed to cover the average heating needs of communal buildings in a typical village in the Covasna region. This small-scale installation can be linked to a district heating network. The necessary area for bioresources needed to meet that capacity is around three hectares; however, the full heat independency would need an area that is about two times larger so as to accommodate peaks in the demand. Each installation needs 60–75 tons of bioresources per year, equivalent to 350–430 m^3 based on an average density of 175–350 kg per m^3. Experiments suggest that the annual consumption of 400–450 m^3 of wood chips ensure heating of 2200 m^2 of communal buildings.

7.4 Conclusions

Short rotation cultivation based on fast growing plants delivers high yield, which can be dried and chipped at a low cost for energy use in buildings. These bioresources provide cost-effective heating based on renewable energy in rural areas. Compared to agriculture, only a few inputs and marginal land is used, whereas plant residues enrich soil and store noxious compounds, such as carbon dioxide. Boilers in the centrally located communal buildings are the starting points for distribution, to houses in district heating networks based on the bioresources. This is economically feasible. Communities can own the networks and willow plantations on the marginal lands, including the essential technologies for wood logging and chipping entailing development into higher value products, based on biorefineries.

References

Aebiom A (2015) Aebiom Statistical Report 2014. Available at www.aebiom.org/blog/aebiom-statisticalreport-2014-2/. Accessed 25 Jan 2017

Bartha S, Vajda B, Duarte L, Carvalheiro C, Antal N (2017) Environmental and economical assessment analysis of the energy willow. A New Biomass Source for the Biorefinery Industry, Resilient Society—Multidisciplinary Contributions from Economic, Law, Policy, Engineering, Agriculture and Life Sciences Fields-Les Presses Agronomiques de Gembloux, Belgium, Bioflux, Romania, pp 9–23

Distribution of land and forest areas amongst member states in the EU27 (Source: FAO, 2010)

HG Order 386/2016-Romanian normative for the thermo technical calculation of buildings

Hart JF (2001) Half a century of cropland change. Geogr Rev 91:525–543. https://doi.org/10.1111/j.1931-0846.2001.tb00239.x

Heimlich RE (1989) Productivity and erodibility of U.S. cropland. Agricultural Economic Report 604, U.S. Dept. of Agriculture, Economic Research Service

http://adevarulfinanciar.ro/articol/romania-din-statistica-fond-forestier-in-crestere-masa-lemnoasa-in-scadere. Accessed at April 2017

Klass DL (1998) Biomass for renewable energy fuels and chemicals. Academic Press, Cambridge

Lal R (2004) Soil carbon sequestration to mitigate climate change. Geoderma 123:1–22

Nonhebel S (1997) Harvesting the sun's energy using agro-ecosystems. Quantitative approaches in system analysis No. 13, Dept. of Theoretical Production Ecology, AB-DLO, Wageningen

Popescu A (2009) Aspecte teoretice si metodologice ale sistemului de indicatori de evidenta, analiza si previziune in silvicultura si economia forestiera An. Inst. de Ist. "G. Bariţiu" din Cluj-Napoca, Series Humanistica, tom. VII, 2009, p. 281–306

Porter ME (1985) Competitive advantage: creating and sustaining superior performance. New York, Free Press. London, Collier Macmillan, 500–560 pp

SR-EN ISO 17225-4, 20140. Solid biofuels—fuels specification and classes
Scarlat N, Bluidea V, Dallemed JF (2011) Assessment of the availability of agricultural and forest residues for bioenergy production in Romania. Biomass Bioenerg 35:1995–2005
Teagasc AFBI (2010) Short rotation coppice willow, Best Practice Guidelines. Available at http://www.seai.ie/Renewables/Bioenergy/Willow_Best_Practice_Guide_2010.pdf
Toivonen MR, Tahavanainen LJ (1998) Profitability of willow cultivation for energy production in Finland. Biomass Bioenerg 15(1):27–37
UNI CEN/TS 14961 (2005) Solid biofuels—fuel specifications and classes, http://www.cen.eu

Chapter 8
District Heating in Villages

Sebestyén Tihamer

The bioresources for energy production in rural areas are available in many countries, but citizens' participation is often missing. District heating based on local bioresources is developed in seven villages in Croatia, Macedonia, Slovenia, Serbia, and Romania. Citizens' participation is fostered, development of district heating based on bioresources is assessed, and the balance of energy demands and supplies in a village is estimated.

8.1 Citizens' Involvement

In many countries, district heating suffers from obsolete infrastructure, equipment, and billing systems (David 2016; Leca 2015). The decline of district heating goes on although expert and policies in the European Union advocate more district heating based on bioresources because they are considered to be most efficient, low carbon supply of heat (EC 2016). Ditrict heating is efficient when the energy-efficient technologies are used (Kim et al. 2017; Lund and Dyrelund 2010; Lund and Vad 2014).

For many inhabitants of rural areas, however, it is cheaper to collect firewood from the forest without any administrative controls on the logging of trees; then dry it in backyards until the next heating season for use in their fireplaces. This practice is widespread, though conventional heaters cause much smoke and smog in the winter and rarely do they achieve more than 40% energy efficiency. Since 2015, legislations in Romania created barriers for the collection of wood. In addition, the growing industrial demand for wood consequently doubled the price of fuelwood within three

S. Tihamer (✉)
Green Energy Innovative Biomass Cluster, Presei Street 4, InnoHUB,
520064 Sfantu Gheorghe, Romania
e-mail: sebestyen.tihamer90@gmail.com

Y. Krozer and M. Narodoslawsky (eds.), *Economics of Bioresources*,
https://doi.org/10.1007/978-3-030-14618-4_8

years. The increase in prices triggered greater interest in the modernization of district heating.

The modernization of heating involves a change in technologies and social relations (Cebotari and Benedek 2017), as well as new spatial behavior because the local supply of bioresources needs much more preparation (Bridge et al. 2013). In addition, utilization of local renewable energy potentials in peripheral rural areas can support the balanced rural development (Benedek et al. 2018). In this perspective, preparations of the modern district heating have been initiated in seven villages across five countries in South East Europe. Villages are selected from those that have shown interest, expressed by its citizens and authorities. The locations of villages are shown in Fig. 8.1.

Fig. 8.1 Locations of villages in the assessment

8.1.1 Citizens' Responses

Based on interviews, citizens' awareness about bioresources and renewable energy was assessed along with their willingness to cooperate and expectations about shifting from the collection of firewood to district heating biomass. Citizens and stakeholders were interviewed using questionnaires in local languages, with 19 questions in two sections. The general section addressed the heat demand of households, construction year, materials of buildings, insulation, and present energy consumption. Another section was about the citizens' satisfaction with the existing heating, willingness to shift to district heating, stakeholder expectations and knowledge, as well as the expected barriers for the use of renewable energy resources. In total, 299 filled out questionnaires were received from citizens and stakeholders. Thereafter, information days were organized by local authorities.

Based on the inquiries, it is found that people are rarely aware about renewable energy. The awareness in Slovenian and Croatian villages was 12% higher than in Romanian villages, where only 22% of the inhabitants had any knowledge about the subject. Nevertheless, more than 65% of the citizens and stakeholders wanted the district heating based on local bioresources, because the existing heating equipment installed is outdated. The shift from fossil fuels to low carbon solutions was welcome. More than 75% of the citizens wanted efficient technologies, while the same percentage expressed a high interest in bioresources. They expected space and water heating. Not many people mentioned job creation, local economy, CO_2 reduction, or creation of new businesses emerging from the local bioresources. A lack of funds and poor technology are considered reasons for the low use of bioresources. The low use of bioresource in Romania was attributed to cheap fuelwood; but ever since the prices tripled because of national regulations, citizens without gas pipelines face difficulties with heating. Reasons cited varied. Lack of technology and financial mechanisms are mentioned by Serbian and Romanian citizens; lack of support from the local authorities say Macedonian respondents; and a lack of local initiatives is mentioned by those interviewed in the Croatian villages.

Most citizens want lower heating costs due to district heating based on bioresources; other issues are less important. They want a connection to the grid of district heating if the price goes down; only 20% are not willing to be connected. The expectations vary across countries: 76% of respondents across all countries expect lower costs of the heating, but 85% in Romanian and 40% in the Slovenian village do not. In Croatia, Slovenia, and Serbia, 15–20% of the citizens are willing to pay more, but this is a negligible number in Romania. In Macedonia and Romania, it has been emphasized that district heating should be accessible to those who fall in lower income categories. People also expect support from national funds or the EU. They welcome ecological technologies as long as they do not entail higher costs.

Most citizens wish for local ownership of bioresources for district heating, and they want this heating for private or public places. Some citizens mention the local authorities, institutions, population, or non-governmental organizations as possible owners. Whether or not they want to be personally involved is a question that was answered only by 21% respondents (63 citizens). About 50% of all respondents want local authorities in the lead, particularly the mayor. About 35% of the respondents from the Slovenian village advocate citizens' ownership, while the Serbian and Romanian villagers rarely mention this option for ownership. Slovenians underline importance of information about community ownership from the neighboring Austria. Others do not mention this option, but they want municipal ownership in co-operation with public utility or public–private partnerships. Only a few respondents point to private companies. Many villagers consider heating as a social service to be delivered by a public organization. It is often argued that entrepreneurs in their community are not capable of delivering district heating. Only a few respondents expect a business opportunity in district heating in rural areas.

Most respondents argued that all relevant stakeholders should be involved in the planning and implementation of district heating, including those citizens who will become consumers in the future. The most suitable public and private buildings are pinpointed. Many respond that they have little knowledge about district heating, though 71% of the respondents answered positively about their participation demonstrating their support for the idea. As many as 61% of the respondents want to participate in working groups and meetings, several are ready to participate. The citizens' support for district heating based on bioresources is assessed as sufficient, and there exists interest for the participation.

8.1.2 Citizens' Participation

The participation of citizens is fostered by introducing the village representatives and know-how in Germany and Austria, because in these countries exist the most suitable experiences with district heating based on bioresources. Tours were organized to Bavaria and Baden-Würtenberg in Germany, as well as to Styria and Burgenland in Austria. The cultivation, storage, and management of bioresources, as well as production and use of biofuels and heat distribution were demonstrated.

In the villages, German and Austrian experts were invited for capacity building. Several seminars and workshops were organized with the German and Austrian experts who shared their experiences regarding the development and implementation of bioenergy villages. Points of information were established in each village, intended to function as a helpdesk for citizens during work hours—offering information, advice, and consultancy at no cost. These are located at the local council, in schools, or at offices of public administration. The Slovenian helpdesk received 2–3 enquiries per week; the Croatian village Perušić 10 per month; Lekenik 8 per month; along with phone calls, the Romanian village Estelic received 10 enquiries per month; Ghelinta 12 per month, whereas the Serbia helpdesk got 20 inquiries in two years and the Macedonian one reached 104 households in 2 years. Mayors of the villages expressed their support, and members of local councils, directors of schools and such authorities answered citizens' questions.

Local working groups were set up in each village with representatives of the local administration, school directors, teachers, local volunteers, members of local councils, managers of public utility companies, district heating networks, regional development organizations and national parks, as well as foresters, farmers, and national experts. They met for deliberations of strategies, preparation of projects, and action plans. They also outlined long-term strategies, worked on training and dissemination of information in schools, culture centers, and other public events.

In addition, between 2016 and 2018, more than 15,000 citizens participated in the public events held in those seven villages. The kick-off usually was the opening of the information helpdesk, followed events broadcasted on television and published in the regional papers. Progress was monitored through monthly presentations to the public in said villages. The promotional campaign was intensive and effective as high participation was achieved.

8.2 Feasibility

The feasibility of biomass production, storage, and processing for district heating is assessed by the Austrian and German experts based on local surveys. The results were positive when the scale of production was in tune with demands. Appendix presents basic data: Table 8.1 contains the basic data, Table 8.2 contains the economic results, and Table 8.3 contains the social and environmental impacts.

8.2.1 Assessments

The number of customers served with biomass based heating varies from 13 to 22 per village, primarily comprising of public buildings. A customer can be a school, kindergarten, or administrative building where many people stay, which operates as the central point in the distribution to households in the village. Full implementation of the biomass district heating system would serve more than 1000 local inhabitants, which is more than 20% of the local population. Investments in the installations reflect the scale of district heating and vary from €0.415 million to €5.3 million. Employing the 25 years of depreciation and realistic energy prices for the income from heat sales, the profitability measured by the internal rate of return varies across villages from 7.1% to 11.9%. If all villages install district heating based on the bioresources, a minimum of 51 new jobs per year can be created and 2110 ton CO_2 per year would be reduced. In the largest village, which is Perušić in Croatia, 2 MW_e and 8 MW_{th} plants for combined heat and power (CHP) are envisaged for—electricity sales to the grid and distribution of heat to 202 customers. While the investments are feasible, not all are commercially attractive. Local working groups found these proposals to be viable and attractive for the identification of investors, communication with customers and surveying of heat demands. The technical plan and detailed costing were prepared in a few villages.

8.2.2 *Value Chain*

Romania is a forest rich country with 1.24 ha forest per capita, four times the EU-28 average of 0.35 ha/capita. Furthermore, Romania possesses a standing stock of about 212 m^3 timber per ha compared to 163 m^3 in the EU-28. The bioresources and biofuels value chain in the municipality of Ghelinţa in Romania was assessed. According to the latest census in the Municipality of Ghelința, 5191 inhabitants reside there in 2018 and this number grows. Ghelinţa is a bioresources-rich municipality with several logging companies that provide income to about 42% of all its citizens, employed in more than 20 sawmills in the hands of four owners or associations.

The bioresources can be generated from forests in the municipal area of 6430 ha, about half of which rests in private ownership. Most trees are older than 80 years of age, and the afforestation on several hundred hectares covers young trees. Spruce and pine dominate the mountains, beech, oak, and poplar the lower altitudes. Agriculture on 3764 ha is extensive and done in small-scale units because farms are an average of 4 ha. They produce mainly cereals on 1130 ha as well as corn, potatoes, sugar beets, and vegetables.

About 10,000 m^3 trees are logged per year. Wood waste from forest cleaning and thinning exceeds 5300 tons a year, which provides potentially 27 GWh per year. The wood-based industry in Ghelința produces about 1600 tons residues which have the potential of 6 GWh per year. Energy from the agricultural residues, from orchards and plantations on marginal lands, after the subtraction for higher value uses such as fodder, is potentially about 12 GWh a year. Livestock excrements from chicken, cattle, sheep, and pigs, if bulked, are potentially 5 GWh per year. All bioresources in the municipality constitute a potential of 50 GWh a year. Other scattered bioresources add to that as well. For example, the municipality is responsible for maintaining local public areas, as well as more than 500 ha montainous pastures and several riverbeds. There exist other small-scale private bioresources, such as food residues.

Slated for 2019, a biomass storage center for wood, dust, and biomass residues from backyards, gardens, and the like is being planned on 2500 m^2. It includes a roofed biomass deposit center with buildings in four sub-divisions on 1500 m^2 which would have 6000 m^3 capacity. Biofuel would be stored on 721 m^3 during the full scale of operations of district heating for 10 h a day, spanning 20 days. The municipality also aims to establish a local bioresources center with a new tractor, trailers, and a wood chipper. These costs of storage and machines are estimated to be an additional € 300,000 in total.

The municipal plan is to connect buildings within a short distance from the Ghelința center. The aim is to construct in three phases: first, the public building connection to the heat grid; then, small businesses, stores, flats and finally to connect a maximum of 52 households on neighboring streets. The plan is purposed to support 12 public buildings, three blocks of flats and four private buildings for households on about 10,000 m^2, which can meet the demand of 2732 MWh per year heat. An additional 492 MWh heat demand has been identified in three blocks of flats based on the measurement of flats, energy classification of the buildings, length of the heating season and refurbishment plans, but the owners have not shown interest. Instead, a few shops and a guesthouse are included. The bioresources available within the municipal borders exceed the demands of the total village by 20 times, which is sufficient even under conservative assumptions about energy efficiency.

Assuming a 16% overall loss in the grids, the total boiler capacity should be 1.27 MW to be delivered by three biomass boilers of 0.9 MW each, with an 83% average boiler energy efficiency. A backup boiler of 1 MW can supply 0.47 MW for peak load, which is sufficient to cover the maximum heat demand in the case that a boiler is down. The existing oil boiler will produce about 6% of the heat demanded and will be used for a few days in a year, as a backup during times of peak load.

The boilers of that capacity need an investment of € 705,000. Based on € 15 per MWh and the price of wood chips, an average sales price of heat at € 61.5 per MWh is envisaged. The payback time of this investment is 12.4 years; which is a feasible investment but commercially unattractive. Laying out the grid of 1500 m for all three phases with pipe and earthwork costs € 300,000, and the installation of the energy transfer stations € 50,000. These depreciate in 14 years. The total investment is about € 1.1 million and the payback about 14.2 years.

The costs of the Ghelința district heating system based on bioresources are assessed, meaning the costs per energy unit after correction for inflation, interest rate, and work hours. The levelized costs of district heating are € 56.9 per MWh. The purchase of bioresources is the largest cost factor throughout the life span. Assuming the sales price of € 61.5 per MWh heat, which increases by 2.2% per year during the 25 years, the net present value is about € 152,000 without subsidies. It implies an 8.4% internal rate of return. If local heat and power supply can cover up to 98% of all investment costs for this non-profit service to local public institutions, the Ghelința authority intends and aims to provide a national subsidy for modernization. Details regarding implementation can be shared upon request.

8.3 Conclusions

There is little awareness but high interest for district heating based on bioresources in villages of South East Europe. Although the villagers' knowledge about such systems is low, many are willing to participate. The main motivation is lower costs of heating, because not many expect many jobs and businesses from its implementation. Authorities and citizens can also participate in decision making and contribute to the implementation when they are equipped with reliable information. District heating is assessed to be cost-effective to citizens and feasible, but commercially not attractive because the profitability of investment is low. Assessment of the value chain for a village, illustrative of many other bioresources-rich villages, shows sufficient organic residues to be able to provide for a large part of the heating demands (in the village), at a reasonable price. Nonetheless, assessments can be biased toward district heating based on bioresources because motivated villages are selected.

Appendix

See Tables 8.1, 8.2 and 8.3.

Table 8.1 Technical parameters for the accounts. *Source* Biovill.eu

Technical parameters	Dole pri Litiji, Slovenia	Estelnic, Romania	Ghelinta, Romania	Kichevo, Macedonia	Kostojevići, Serbia	Lekenik, Croatia	Perušić, Croatia
Consumer No/DH grid length (m)	18/890	20/678	41/1.500	13/1.220	93/2000	18/1.126	202/9.084
District Heat (DH) sale (MWh/a)	493	5210	2.206	8.842	1.870	1.088	12.367
Network heat utilization ratio (kWh/m/a)	554	5568	1.470	7.247	1.456	966	1.360
Total consumer peak H load (MW)	0.38	2.84	1.1	6.8	1.4	0.84	8.25
Grid heat losses (% H prod.)	21	14	16	10	25	13	18
Peak load (Boiler fringes) (MW)	0.45	2.2	1.04	7.25	1.12	0.9	5.03
Total biomass boiler capacity (MW)	0.45 MW	1.85	0.7	5	0.7	0.6	2.4 + 1
Backup boiler capacity (MW)	Not done	2.5	0.65	3	0.75	0.6	4
Biomass demand (MWh/a)	752	6.942	2.974	11.363	2.734	1.492	17.080
Biomass energy fraction of DH (%)	100	96	94	96	91	99	94
Wood-chip demand (w40, P31) (m^3/a)	820	1.735	1.126	10.230	1.012	1.835	13.596
Full-load hours of biomass boilers (h/a)	1.390	2.880	3.526	1.886	3.241	2.060	4.170
Plant efficiency (DH sales/fuel input) (%)	65.5	75	69.5	75	62	72	68

Table 8.2 Economic parameters of investments in target villages. *Source* Biovill.eu

Parameters, costs in €	Dole pri Litiji, Slovenia	Estelnic Romania	Ghelinta, Romania	Kichevo, Macedonia	Kostojevići Serbia	Lekenik Croatia	Perušić Croatia
Service life (a)	25	25	25	25	25	25	25
Initial investment	415,000	1,020,000	610,000	1,220,000	280,000	594,000	5,343,000
Investment subsidy	214,700	970,000	580,000	0	0	84,000	756,000
Reinvestment 2030	120,000	600,000	204,000	950,000	200,000	262,000	980,000
Biomass price, €/MWh	17.9	15	15	17.5	15.6	18.8	15.75
Biomass purchase	13,400	104,000	44,600	199,000	43,000	28,000	269,000
Heat sale €/MWh	86.5	54.5	56.9	44.9	50.3	82.9	65.4
Heat price €/MWh	90	55.6	61.5	46.5	52	85	66
Income	44,400	284,000	135,000	411,000	96,000	92,500	847,000
NPV	18,900	104,000	152,500	250,000	56,000	39,000	576,000
Internal rate of return	7.3%	11.9%	8.4%	7.1%	8.7%	5.4%	6.4%
Discount payback time	2.5	8.6	14.2	4.5	10.5	23	16.7

Table 8.3 Socio-economic and environmental impacts in target villages. *Source* Biovill.eu

Parameter	Dole pri Litiji, Slovenia	Estelnic, Romania	Ghelinta Romania	Kichevo, Macedonia	Kostojevići, Serbia	Lekenik Croatia	Perušić Croatia
Bioenergy (MWh/a)	752	5210	2974	11,363	2734	1492	17,080
Share of bioenergy (%)	64	100	100	34	100	100	83
Surplus of bioenergy (MWh/a)	−189	50	395	−8842	84.4	12,661	0
Surplus of bioenergy share (%)	0	26	120	0	8	N.a.	−980
Cost saving compared to fossil fuel €	34,900	129,000	55,000	289,000	289,000	27,600	469,500
Reduced CO_2 t/year	221	2032	871	3.334	798	438	4938
New full-time jobs	1	8	4	13	3	2	20
Cost saving €/MWh	36–56	0–8.5	0–4	12	2	2.7	19.2

References

Benedek J, Sebestyen T, Bartok B (2018) Evaluation of renewable energy sources in peripheral areas and renewable energy-based rural development. Renew Sustain Energy Rev 90:516–535

Biovill, Bioenergy Villages (BioVill)—Increasing the Market Uptake of Sustainable Bioenergy. This project has received funding from the European Union's Horizon 2020 research and innovation programme under Grant Agreement N° 691661, http://biovill.eu/de/biovill-publications-en/

Bridge G, Bouzarovski S, Bradshaw M, Eyred N (2013) Geographies of energy transition: space, place and the low-carbon economy. Energy Policy 53:331–340

Cebotari S, Benedek J (2017) Renewable energy project as a source of innovation in rural communities: lessons from the periphery. Sustainability 9, 509. https://doi.org/10.3390/su9040509

David A (2016) The development of the urban heating sector in Romania. Aalborg University Copenhagen. http://projekter.aau.dk/projekter/files/239538001/Master_thesis_final.pdf, Accessed: 20 Aug 2017

EC (2016) Energy for all Europeans, communication from the commission to the European parliament, Brussels, 30 Nov 2016 pp 1–13

Kalkbrenner BJ, Roosen J (2016) Citizens' willingness to participate in local renewable energy projects: The role of community and trust in Germany. Energy Res Soc Sci 13:60–70

Kim HJ, Lim SY, Yoo SH (2017) The convenience benefits of the district heating system over individual heating systems in Korean households. Sustainability 9:1384 (10,3390/su9081348)

Leca A (2015) Romania needs a strategy for thermal energy. Manag Mark 10(1):3–11 (10.1515/mmcks)

Lund H, Dyrelund A (2010) The role of district heating in future renewable energy systems. Energy 35(3):1381–1390

Lund H, Vad MB (2014) 4th generation distirct heating (4GDH): integrating smart thermal grids into future sustainable energy systems. Energy 68:1–11

Chapter 9
Renewable-Based Gas Systems in Austria

Karin Fazeni-Fraisl, Johannes Lindorfer and Robert Tichler

This chapter is about the potential production volumes of sustainably produced, bio-based methane (CH_4) from anaerobic digestion, and synthetic natural gas. It refers to power-to-gas processes. Though based on the ground reality in Austria, the insights and learnings are useful for other countries too. The chapter starts with a characterization of available bioresources for anaerobic digestion, which is used as a basis for estimating the total theoretical potential of biogenic CH_4 for the heating sector. Furthermore, the potential of synthetic natural gas in Austria has been estimated based on a projected increase of power generation. Wind and photovoltaics are used for producing H_2 through the process of electrolysis. Another system component for H_2 methanation is CO_2 from biogas purification. The overall aim is to show how, and at what costs, renewable gases based on renewable resources can contribute to an energy supply for the heating sector.

9.1 Introduction

In 2016, the supply of natural gas was approximately 17% of the end user energy demand in Austria. Natural gas played an important role in the heating sector, where it covered about 25% of the end user demand for energy, equivalent to about PJ 77.3. In addition, that year, district heating covered about 20% of the total demand of end user energy for heating, which was PJ 60.3 (Statistik Austria 2017). Natural gas covered 39% of all energy resources used for district heating; however, more than half of the energy resources used for district heating that year were renewables such as wood biomass, garbage, biogas, landfill gas, and industrial waste heat (FGW 2017).

K. Fazeni-Fraisl (✉) · J. Lindorfer · R. Tichler
Energy Institute at the Johannes Kepler University, Linz, Austria
e-mail: fazeni@energieinstitut-linz.at

Y. Krozer and M. Narodoslawsky (eds.), *Economics of Bioresources*,
https://doi.org/10.1007/978-3-030-14618-4_9

These numbers indicated the dependence of heating on natural gas and renewable energy in Austria. Presently, the two major drivers for additional renewable energy are: one, the EU target of a 40% reduction (compared to 1990 levels) in greenhouse gases emissions by 2030[1]; two, the proposal in the revised Renewable Energy Directive that calls for a one percent annual increase of renewables in the share of energy resources used for heating.[2]

The two options are the integration of methane (CH_4) from the anaerobic digestion of bioresources as well as the integration of synthetic natural gas produced in a power-to-gas process. Synthetic natural gas is produced via a methanation process, which uses hydrogen (H_2) from electrolysis, and a carbon source such as carbon dioxide (CO_2). As for the distribution of biogas and synthetic gas based on renewable resources, it makes sense to use the existing infrastructure in place for natural gas, at least as long as it is cheaper than other options. The power-to-gas system enhances not only the integration of a renewable energy carrier into the existing natural gas grid, but it is also considered an important building block for integrating volatile electricity generation based on wind or photovoltaics into the energy system. It is a step forward in linking the electricity and gas grids. For a sustainable gas supply in the future, it is important to integrate CH_4 from anaerobic digestion and power-to-gas. Striving for synergies between these systems is essential; for example, using CO_2 from biogas purification for the methanation process, thereby storing renewables-based surplus of electricity in the gas grid.

9.2 Bioresources Potential for Biomethane

Competition between agricultural production for food, and renewable energy resources is intensively discussed and options for sustainable energy supplies are studied (Nonhebel 2007; Koh und Ghazoul 2008; Alexandratos 2008; Escobar et al. 2009; Tilman et al. 2009; Franco et al. 2010; Rathmann et al. 2010; Karp and Richter 2011). In order to ensure a sustainable future for gas supply in Austria, only biogenic waste materials are considered as feedstock for anaerobic digestion that produces biogenic CH_4.

Agriculture Residues

The bioresource potential from agricultural residues is estimated for the year 2015, then extrapolated to the years 2030 and 2050 based on projected population growth and economic development in the agricultural sector; for instance the projections in animal husbandry and arable land. The result is a time series for theoretical biogenic waste material available for biomethane production in Austria. Based on that, the

[1]European Council 23/24 October 2014—Conclusions: http://www.consilium.europa.eu/uedocs/cms_data/docs/pressdata/en/ec/145397.pdf.

[2]Proposal for a DIRECTIVE OF THE EUROPEAN PARLIAMENT AND OF THE COUNCIL on the Promotion of the Use of Energy from Renewable Sources (recast) COM/2016/0767 final/2-2016/0382 (COD).

total biomethane potential for injection into the existing natural gas grid is estimated. The estimates are theoretical potentials because limits posed by the competing uses of biomethane, or logistics in the supply of the biogenic waste material are assumed solvable. Agricultural harvest is an exception as not all straw can be harvested since it is required for animal husbandry and maintaining soil fertility. Therefore, it is assumed that about 40% of the total cereal straw harvest is usable for anaerobic digestion. For corn stover and rapeseed straw, it is assumed that 100% of the harvestable amount can be used for energy because there are no competing demands from animal husbandry. Further, it was assumed that 100% of turnip leaves are left in the field. As the current practice is plowing into the soil (Vetter et al. 2012), there is a high potential for turnip leaves to be used as a feedstock for biogas. Due to plowing of the turnip leaves, the nutrients must be replaced by spreading of digestate from the anaerobic digestion of manure.

In general, harvest residues that can be sustainably used for energetic purposes differ regionally depending on farming practices, as well as soil and climatic conditions (FAO 2012; Blanco-Canqui 2013; Karlen et al. 2014). The yield of agricultural residues is estimated using a literature-based average about corn-straw ratios and the Austrian harvest statistics for 2015 (Bundesanstalt für Agrarwirtschaft 2016). The projection of agricultural residue potentials until the year 2050 is based on the development of cereal and rapeseed, as well as on the area of rapeseed planting in Austria from 1990 until 2015 (Bundesanstalt für Agrarwirtschaft 2016). Linear regression based on historical data is used to predict the area of arable land in the future, up to 2050. A long-term decrease in the planting areas of cereal, rapeseed, and sugar beet was observed. For projecting estimates between 2016 and 2050, an annual decrease of 0.4% in cereal acreage is assumed until 2041 and 0.5% from the year 2042 onwards; for rapeseed by 0.9% from 2016 to 2034, then 1% until 2034 and 1.1% from 2045 onwards; and for sugar beet by 0.6% from 2016 to 2024, 0.7% from 2025 to 2045 and 0.8% from 2046 to 2050.

In addition, manure and slurry from animal husbandry are suitable feedstocks for biomethane production. The Environment Agency Austria developed two assessments for using manure and slurry as biogas substrates (Zethner und Süßenbacher 2012):

1. The utilization of suitable untreated manure and slurry delivered by all Austrian animal husbandry. This amounts to 25 million tons per year, and comprises 45% of all manure; 29% of all manure slurry comes from cattle farming (6% of all manure), 80% of all slurry from pig farming and 100% from horse and poultry manures.
2. The utilization of slurry and manure only from farms larger than 50 livestock units forms the potential for anaerobic digestion at about 10 million tons per year, which covers about 31% of total Austrian slurry and manure production.

The second assessment is used for biomethane potentials, which means 10 million tons of slurry and manure per year. Using that, the development of biomethane was projected till 2050 by extrapolating historical data of Austrian animal husbandry from 1999 to 2015 (Bundesanstalt für Agrarwirtschaft 2016). It shows a steady decrease of cattle and pig manure, but an increase in that of chicken. Since the number of horses fluctuated per year, the utilization of manure is assumed to be constant. The total manure and slurry is scaled in proportion to the share of each animal group in Austrian farming (Bundesanstalt für Agrarwirtschaft 2016).

The Food Industry

Residues from the food industry are assessed based on the latest data from 2009 to 2012. In Austria, about 1.3 million tons of biogenic residues from the food industry are found (Reisinger et al. 2012). In addition, from meat and dairy processing residues (FABbiogas n.a.), about 2.3 million tons of biogenic waste material is available for biomethane production in Austria; assumed constant from 2016 to 2050.

Biogenic Waste

Biogenic waste covers green cuttings, organic waste from households disposed in the organic waste bin, and food residues disposed in the residual waste bin of a household. In Austria, green cuttings are estimated at 47 kg per capita per year from private gardens, and 54 kg per capita per year from public areas (BMLFUW 2015). Food waste in municipal, household, and retail waste is estimated at 77 kg per capita per year. In addition, it is estimated that 177 kg per capita per year of biogenic waste is composted in house gardens (Pladerer et al. 2016; Scherhaufer et al. 2016). Based on the Austrian demographic, an increase in biogenic waste material is projected up to 2050, but this excludes the component of composting in private gardens, which is assumed to remain constant.

Sewage

In Austria, the annual sludge from waste water treatment plants > $50EW_{60}$ is estimated at 31 kg per capita (BMLFUW 2014). The projection is also based on demographic data up to the year 2050. Table 9.1 shows theoretical potentials of biogenic waste material in Austria for biomethane production.

Slurry and manure will make up most of the potential, estimated at 3.3 and 3.4 million tons by the year 2050, followed by biogenic waste material from garden composting about 1.5 million tons and food industry waste of about 1.3 million tons. In contrast to agricultural waste material which declines over time due to declining livestock and arable land, biogenic waste increases due to population growth, whereas waste from the food industry is assumed constant due to a lack of valid data for projection estimates. The feedstock potentials presented in Table 9.2 are used to calculate the bioemthane production potential.

Resource Potential for Synthetic Gas

In general, CO_2 is not a limiting factor for power-to-gas processes because it is available from fermentation and digestion processes, as well as from the combustion of fossil fuels and chemical synthesis. The production of synthetic natural gas is limited by a surplus of electricity that is needed for the production of hydrogen (H_2) with electrolysis. The electricity surplus in Austria, based on the growth of hydro, solar, and wind power, is estimated assuming moderate growth in Austrian policy projections. Electricity produced from non-renewable resources is excluded because it generates CO_2 and therefore does not contribute to a low-carbon future of the gas system (Reiter and Lindorfer 2015). That surplus is defined by those scenarios as the electricity production that exceeds demand. Nonetheless, there exists a high degree of uncertainty in the projections made up to 2050 because the surplus is dependent on the growth of renewable electricity production. The scenarios are shown in Table 9.2. The following assumptions are made to quantify the theoretical H_2 production potential, and downstream, the CH_4 potential (Reiter and Lindorfer 2015):

- Full capacity utilization of electrolysis and methanation
- Electricity input for electrolysis: 5 kWh/m^3 H_2
- H_2 input for methanation: 0.5 kg H_2/kg CH_4
- CO_2 input for methanation: 2.75 kg CO_2/kg CH_4
- Full load hours electrolysis: 2500 h/a
- Full load hours methanation: 2500 h/a.

Surplus electricity for electrolysis-based hydrogen production grows until 2050. For the estimation of the synthetic natural gas, the hydrogen potential is needed. Based on those assumptions, the theoretical H_2 production is potentially 5657 GWha^{-1} in 2030 and 7044 GWha^{-1} in 2050.

Table 9.1 Theoretical potentials of biogenic waste material in Austria for biomethane production

Ton/a	2015	2020	2030	2040	2050
Slurry	5,709,928	5,352,680	4,709,635	3,995,140	3,282,716
Manure	4,390,072	4,237,589	3,962,269	3,657,302	3,352,753
Cereal straw	1,034,369	1,013,414	975,694	933,784	891,893
Corn stover	655,162	641,889	617,998	591,452	564,918
Rapeseed straw	58,166	55,802	51,547	46,820	42,097
Turnip leaves	453,820	439,766	414,468	386,360	358,273
Plant based food industry waste	1,254,610	1,254,610	1,254,610	1,254,610	1,254,610
Animal by-products	966,700	966,700	966,700	966,700	966,700
Food waste in residual waste	666,127	695,560	728,078	744,654	754,252
Biogenic waste bin	524,000	547,153	572,733	585,772	593,323
Kitchen and food waste	104,700	109,326	114,437	117,043	118,551
Green cut material	879,300	918,152	961,077	982,956	995,627
Garden composting	1,527,425	1,527,425	1,527,425	1,527,425	1,527,425
Sewage sludge	266,306	278,073	291,073	297,700	301,537

Source Energieinstitut an der Johannes Kepler Universität

Table 9.2 Projection of renewable electricity production in GWh from wind power, photovoltaic, and hydropower for H_2 production

Year	2015	2020	2030	2040	2050
Hydropower	37,656	40,681	41,433	41,433	41,433
Solar power	1478	4810	11,189	17,071	17,771
Wind power	4571	6249	10,785	13,078	15,628
Surplus—10% of total hydropower production	3766	4068	4143	4143	4143
Surplus—15% of total solar power production	222	722	1678	2561	2666
Surplus—20% of total wind power production	914	1250	2157	2616	3126

Source Energieinstitut an der Johannes Kepler Universität based on Österreichische Energeiagentur (2015)

9.3 Biomethane Potential

The potential of biogenic CH_4 for heating and cooling in Austria is estimated, and the following assumptions have been made:

- The existing capacity of biomethane injection into the grid in Austria is about 3000 Nm^3/h, which corresponds to 24 million Nm3 biomethane per year at 8000 full load hours; the biomethane contribution is assumed to be in addition to the existing capacity.
- Biomethane contribution is assumed to start in 2019 with an injection of 1% of the total estimated potential. From 2020 to 2050, an annual growth rate of 3% is assumed, aiming at having the total potential realized in the year 2050.
- In the first stage (2019–2030), anaerobic digestion of plants is used to ferment biogenic waste from households, the food industry, retailers, and sewage sludge. This is done in small plants (50–200 Nm^3/h), and in medium-scale plants (200–500 Nm^3/h) from 2025 onwards. In the second stage (2030–2050), it is scaled up based on the fermentation of agricultural residues such as straw and manure. Hence, highly flexible, multi-feedstock plants (500–1000 Nm3/h) are assumed.

The biomethane potential is compared with the projected end user energy demand from Austrian households for natural gas, and with the scenarios that are developed by the Austrian Environment Agency (the WEM scenario with existing measures and the WAM scenario with additional measures, Krutzler et al. 2015).

The biomethane potential in 2018 is about 1.57 billion Nm^3/a. Of this total biomethane potential, manure and slurry account for 18%; straw represents about 21%; and the residues from food processing about 17%. Biogenic residues, which are composted by households, account for another 18%. Green cut material, biogenic waste bin material, and sewage sludge account for a smaller proportion of about 5–6%. Taking into consideration the decreasing availability of agricultural residues as we approach 2050, the estimated biomethane potential decreases to 1.50 billion Nm^3/a. Herewith, a significant growth of anaerobic digestion is necessary.

Currently in Austria, there are 13 anaerobic digestion plants that produce biomethane for injection into the grid. By the year 2050, 558 plants would be required to achieve the biomethane production that has been envisaged. About 370 of them would be smaller plants, with a capacity of 50–200 Nm^3/h; and the remaining ones would be large, multi-feedstock plants with large sourcing areas and a capacity of 500–1000 Nm^3/h.

The Synthetic Gas Potential

The synthetic natural gas potential for Austria is estimated based on the following assumptions:

- Mainly small-scale power plants with a capacity of 0.5–1 MW are installed until 2020. From 2020 onwards, medium-scale plants with a capacity of 1–5 MW are also installed. The medium-scale ones cover 10% of the total envisaged capacity of power-to-gas plants. From 2025 onwards, large-scale power plants with a capacity of 5–10 MW are envisaged. The large plants will initially cover about 50% of the total installed capacity, which increases to about 65% from 2040 onwards. It is assumed that the plant capacities increase as the power-to-gas technology matures; due to economies of scale and technological advances.
- The power-to-gas plants are assumed to be installed with a capacity of 500 million Nm^3/a by 2050; the power plants start in the year 2020 with permits, planning, construction, etc.
- The annual rate of capacity increase is 1% from 2020 to 2025; then 2% from 2025 to 2030; and finally, 5% from 2030 to 2050.

Table 9.3 shows the theoretical potential of synthetic natural gas from power-to-gas plants in Austria, based on the above assumptions. The potential of synthetic natural gas is estimated at about 303 million Nm^3 in 2020, which increases to 488 million Nm^3 by 2050. This covers 30 and 35% of the WEM and WAM scenario, respectively, which is equivalent of the total natural gas demand of Austrian households (Krutzler et al. 2015). The realization of this potential starts in the year 2020 in which about 2% of the total potential is realized; by 2030, 24% of the total potential is already realized, and finally, the goal of 100% potential realization is reached in the year 2050. To realize that potential, 1520 power-to-gas plants are needed by 2050, these comprise 1174 small plants with 0.5–1 MW capacity, 117 medium-scale plants with 1–5 MW capacity and 229 larger power-to-gas plants with 5–10 MW capacity.

Table 9.3 Theoretical synthetic methane potential from power-to-gas in Austria

Million [Nm^3/a]	2018	2020	2030	2040	2050
Smaller sized plants 0.5–1 MW	0	50	43.2	93.7	158.0
Medium sized plants 1–5 MW	0	0	9.3	29.6	49.9
Larger sized plants 5–10 MW	0	0	46.3	192.3	324.3
Total potential of synthetic CH_4	280.4	303.3	400.7	468.1	532.2

Source Energieinstitut an der Johannes Kepler Universität Linz

Total Theoretical Potential for Renewable Methane

Table 9.4 shows the potential for 'green methane' in Austria due to the production of biomethane and synthetic natural gas. In the year 2050, the annual production is assumed to approximate 2 billion Nm^3 of renewable methane, which suffices for the total gas demand of households and district heating.

It is a theoretical maximum for Austria, based on an estimated bioresource potential and electricity capacity for hydrogen production between now and 2050. Such an achievement requires changes in economic, legal, and organizational systems in the Austrian energy, agricultural and waste management sectors because the current systems do not support the expansion of renewable gas production. Another challenge is the development of innovative logistics systems in waste management, in particular organizational changes for the collection of biogenic residues from households, industry, canteens, and retail outlets, including changes in regulations toward that purpose. Planning and implementation of dialog with stakeholders such as farmers, waste management companies, municipalities, energy suppliers, and policy makers is needed in order to maximize social acceptance of the changes. Technological hurdles consist of seasonal changes in feedstock compositions, contaminants in organic waste, matching plant scale to the regional feedstock and digestate use in agriculture and composting plants. Summing up is the realization that the estimated potential requires organizational changes as well as technological development. From today's perspectives, these potentials are not fully exploitable.

For synthetic natural gas, the development of renewable electricity is particularly important. The availability of renewable electricity influences the economics of power-to-gas and base load capacity. In addition, storage must grow to cope with the variability of renewable electricity. If the production of biomethane and synthetic

Table 9.4 Total theoretical potential for renewable methane in Austria

Millions [Nm^3/a]	2018	2020	2030	2040	2050
Existing biomethane grid injection 2017	24.0	24.0	24.0	24.0	24.0
Biomethane potential	0	65.1	566.1	1,067.1	1,518.1
Synthetic methane potential	0	5.0	98.8	315.5	498.9
'Green Methane' for Heating and Cooling	**24.0**	**94.1**	**689.0**	**1406.7**	**2041.0**
Total theoretical methane potential for heating and cooling	1852.6	1874.3	1959.1	2000.4	2017.0
Natural gas demand of households and district heating (WEM)	–	2276.5	2191.3	2116.2	1935.8
Natural gas demand of households and district heating (WAM)	–	2079.2	2001.2	1918.3	1754.0
Share of 'Green Methane'; on natural gas demand of households (WEM) [%]	–	**4%**	**31%**	**66%**	**105%**
Share of 'Green Methane' on natural gas demand of households WAM [%]	–	**5%**	**34%**	**73%**	**116%**

Source Energieinstitut an der Johannes Kepler Universität Linz natural gas is coordinated, the availability of CO_2 for methanation should not be problematic because it can be obtained from biogas purification.

As Table 9.4 shows, about 31 and 34% of the total natural gas demand of households can be supplied by 'green methane' in 2030; the percentage depends on the national energy scenario. This share grows up to 105% in 2030 and 116% in the year 2050. In the year 2030, about 40% of the total 'green methane' potential is realized, and by 2050, the goal of 100% potential realization is reached.

9.4 Economics of Renewable Gas Production

Economics of Biomethane Production

The production costs of biomethane based on agricultural and food residues, animal husbandry, and biogenic waste are estimated. These estimates cover the total costs of biomethane production which include equipment costs, interest, and depreciation rates, plus operating and transport costs. The production cost per unit biomethane varies from 6.2 to 9.4 €-cent/kWh, depending on the capacity of the plant. The cost of the biogas plant is a major factor for biomethane production, followed by the cost for upgrading CH_4. In the case of using biogenic waste materials as feedstock, the substrate cost is the least costly item in the biomethane production. As there have only been minor technical advances in biogas plants in the last decade, no cost decrease due to technological progress is assumed (Grond et al. 2013). Figure 9.1 illustrates the economies of scale, in the present and future.

Biomethane production costs are likely to decrease, due to the larger size of anaerobic digestion plants up to 2050. Regarding these economic advantages, the development of large multi-feedstock plants with a scale of 500–1000 Nm^3/h is an attractive way forward.

Economics of Synthetic Natural Gas Production

The total production costs of hydrogen and synthetic natural gas in power-to-gas plants with capacities 0.5–1, 1–5, and 5–10 MW_{el} are estimated. The estimated range of production costs for hydrogen is 7.0–20.7 €-cent/kWh and for synthetic methane 10.5–37.1 €-cent/kWh. Figure 9.2 shows a continuous decrease in production costs, which reflects an economy of scale.

A significant economies-of-scale cost benefit can be observed. Cost reduction is also strongly influenced by the growth of production which facilitates learning by doing. The electrolyzer and methanation unit involve the largest investment costs for a power-to-gas plant. In contrast to biomethane, it is assessed that the production cost for synthetic natural gas will decrease in the future, due to those scale and learning effects. An estimate is that the annual costs of PEM-electrolyzers decrease by 2.2% (Grond et al. 2013), but here 2% is assumed.

Besides capital expenditures, the cost of electricity has a strong influence on the production cost of synthetic natural gas. Cost estimations include the cost for electricity from solar, wind, and hydro power, which are assumed to be about 20€/MWh;

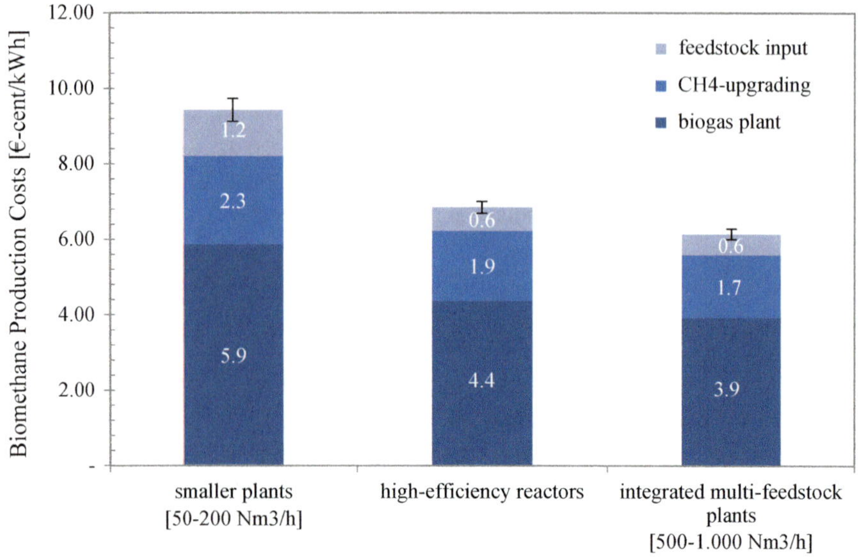

Fig. 9.1 Production costs of biomethane. *Source* Energieinstitut an der Johannes Kepler Universität

a slight increase is expected in the future. It is assumed per year 2500 full load hours of the power-to-gas plant are assumed as well as fluctuating electricity inputs which cause higher expenditures in 2030. Table 9.5 summarizes main economic benchmarks

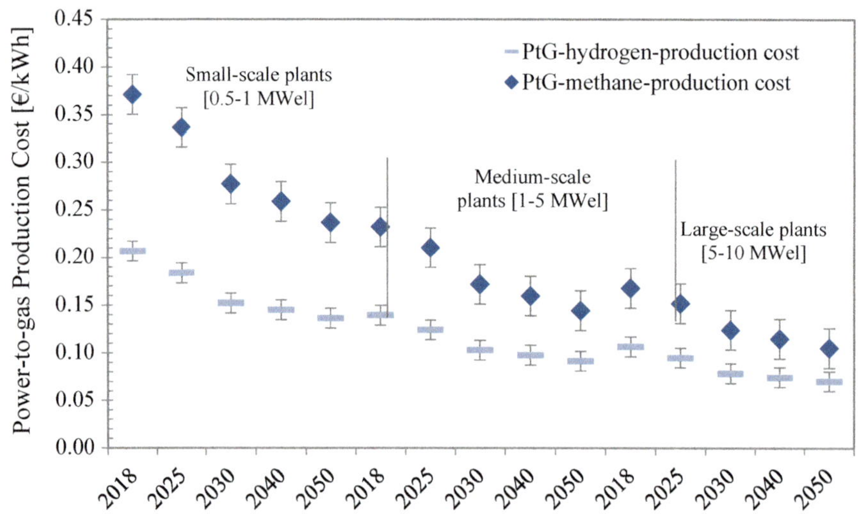

Fig. 9.2 Production cost of hydrogen and synthetic methane in power-to-gas plants. *Source* Energieinstitut an der Johannes Kepler Universität

Table 9.5 Comparison of economic benchmarks

Source	Units	2015		2030	
		Price Excl. Tax	Price Incl. Tax[a]	Price Excl. Tax	Price Incl. Tax
H_2 from steam reforming of natural gas	€-cent/kWh$_{LHV}$	4.8	6.1	6.3	7.6
H_2 from biomass gasification	€-cent/kWh$_{LHV}$	10.3	10.3		
Natural gas	€-cent/kWh$_{LHV}$	3.3	5.8	4.2	6.7
Biomethane	€-cent/kWh$_{LHV}$	7.7	10.2	10.2	12.6

[a]Natural gas levy (0.59 Cent/kWh) and grid use fee for the natural gas grid (1.61 Cent/kWh for industry) in Austria

Source Energieinstitut an der Johannes Kepler Universität—power-to-gas assessment tool PResTiGE

to facilitate a comparison of the production cost of H_2 and CH_4 from a power-to-gas process with the current prices of competing products.

In comparison, H_2 from a power-to-gas process is not yet as competitive as H_2 from natural gas steam reforming. Competitiveness can be achieved by technological advances when large plants are established. Also, when compared to the current prices of natural gas, synthetic methane from a power-to-gas process is not competitive; but they can be competitive when technologies advance and larger plants are installed. Competitive H_2 and CH_4 from power-to-gas plants can be accomplished in the medium to long term, if technologies improve and negative external effects of fossil fuels are internalized.

9.5 Challenges for a Future Gas System Based on Renewables

Feedstock Supply

The potential of bioresources for anaerobic digestion is determined by the maximum amount of feedstock available in Austria. The realization of this potential depends on several factors. One of the most important factors is the design of a logistic chain that enables an energy efficient transport system because such a system supports a positive energy balance and helps in attaining economic feasibility. Logistic chains have to be developed for the collection of feedstock, and its transport to anaerobic digestion facilities. An efficient transport system for the digestate has to be implemented as well.

Storage for seasonal feedstock should be implemented, for instance for turnip leaves and straw. These issues can be solved by technology development in anaerobic digestion to such an extent that the seasonally available feedstock can be processed along with harvesting operations during the harvest season. For straw, low-tech storage solutions are possible, such as stock-piles in the field or nearby the processing plant. Thus, storage would no longer be an obstacle in the future.

Further research on the anaerobic digestion of fast degradable substrates like straw is necessary. Pre-treatment technologies are advancing and being applied on industrial scales for bioethanol production; future developments can modify the pre-treatment of biogas. In addition, solutions are needed to cope with impurities, and the fluctuating composition of feedstock for the anaerobic digestion of biogenic household waste, as well as the food component in municipal waste.

Multi-feedstock plants will form an important component for enhancing the anaerobic digestion of waste also for the biomethane injection to the natural gas grid. The anaerobic digestion that is adaptive to variations in the regional and seasonal feedstock should become a part of the supply chain. The use of regionally available feedstock decreases transport distances. This holds true for the feedstock and digestate. The use of seasonal feedstock also decreases storage, thereby contributing to economic feasibility.

Existing infrastructure cannot achieve the potential demonstrated in this chapter. Technical as well as organizational, legal, and behavioral changes in several sectors are a prerequisite for the enhancement of biomethane production from biogenic waste material, particularly in agriculture and waste management that deliver most of the feedstock for anaerobic digestion. In the agricultural sector, the supply of feedstock requires changes in slurry management, and possibly in soil management, e.g., the application of digestate, extraction of harvest residues, etc. The waste management sector needs to develop new systems of collection for the biogenic waste from households. Legal impediments for the reuse of biogenic waste for biogas should also be resolved.

Systemic Challenges for a Renewables-based Gas System

In Austria, natural gas consumers have to pay gas grid tariffs. These tariffs form an important part of the revenues of gas grid operators and are spent on the maintenance and expansion of the natural gas grids themselves. According to the proposal for a revised Renewable Energy Directive (RED II), the share of renewables in the total end energy of the heating and cooling sector has to increase by 1.3% per year up to 2030 (Epp 2018). Therefore, it is assumed that natural gas for heating and cooling has to be substituted by other energy carriers—and a 'phasing out' could be initiated toward achieving this objective. As a result, more and more customers drop out of the gas market which may incur higher costs of the gas grid. The use of biomethane and synthetic natural gas can bring down infrastructural costs, but the injection of those into the grid is also costly.

High capital costs of the electrolyzer and methanation facilities give rise to high production costs of synthetic methane and H_2, generated from steam reforming or natural gas. Development of biological, in situ methanation can lead to lower investment costs as they do not need chemical-catalytic methanation and gas permeation; a cost reduction of 20% to 25% is possible. The direct utilization of H_2 from power-to-gas plants is cost-effective, but large-scale methanation requires an injection into the Austrian gas grid. The present quality requirements limit it to 4-vol.%.

Nevertheless, power-to-gas technology has several important benefits for a future energy system. One is that, electricity from this variable renewable electricity production can be stored, which optimizes electricity purchasing, performance of electricity generation based on renewable resources, and the seasonal storage of electricity. Two is that, the use of the existing natural gas grid for the transport of energy reduces costs as it avoids the need for a new electricity grid. This is especially relevant to regions with weak electricity grid infrastructure and not much capacity for renewable electricity. Additionally, power-to-gas technology enables the exploitation of renewable electricity production in remote areas through the transport of energy via the natural gas grid. Further, power-to-gas technology also supports the production of renewable products for heating, mobility, and chemicals.

9.6 Conclusions

Austria has a high potential for the production of biomethane from biogenic waste materials, and synthetic natural gas in power-to-gas applications. By 2050, the demand for natural gas from the heating sector could theoretically be met with by renewable gas. Therefore, feedstock based on bioresources and renewable energy should be developed; and this includes innovative logistics for feedstock, regionally adapted plant design, technological development in production facilities, as well as legal and organizational changes. All these are possible and economically feasible, albeit requiring a long-term view on development in the energy sector, and in the feedstock suppliers from agriculture and waste management. One option is for about 2 billion Nm^3 of biogas and synthetic natural gas to meet the entire demand of natural gas from Austrian households in 2050. If economic, the injection of that gas mix into the existing gas grid can be a building block for the renewable energy system.

References

Alexandratos N (2008) Food price surges: Possible causes, past experiences and relevance for exploring long-term prospects. Popul Dev Rev 34(4):663–697

Blanco-Canqui (2013) Crop residue removal for bioenergy reduces soil carbon pools: how can we offset carbon losses. Bioenergy Resour 6:358–371

BMLFUW (2015) Die Bestandsaufnahme der Abfallwirtschaft in Österreich. Statusbericht 2015, Wien 2015

BMLFUW (2014) Kommunales Abwasser. Österreichischer Bericht 2014. Wien 2014

Bundesanstalt für Agrarwirtschaft (2016) Grüner Bericht 2016. http://www.awi.bmlfuw.gv.at/index.php?id=gruenerbericht (dl: 19 Dec 2016)

Epp B (2018) Europe's first-ever renewable heating and cooling target. https://www.solarthermalworld.org/content/europes-first-ever-renewable-heating-and-cooling-target (dl: 03 Sept 2018)

Escobar JC et al (2009) Biofuels: environment, technology and food security. Renew Sustain Energy Rev 13:1275–1287

FABbiogas(n.a.) Biogas production and biogas potentials from residues of the European food and beverage industry. Austria-National Situation. http://www.fabbiogas.eu/fileadmin/user_upload/Download/National_Report_AUSTIRA_german.pdf (dl: 20 Dec 2016)

FAO (2012) Soil organic carbon accumulation and greenhouse gas emission reductions from conservation agriculture: a literature review. Integr Crop Manage 16-2012, Rome

FGW (2017) Zahlenspiegel. Erdgas und Fernwärme in Österreich, Wien, p 2017

Franco J et al (2010) Assumptions in the European Union biofuels policy: frictions with experiences in Germany, Brazil and Mozambique. J Peasant Stud 37(4):661–698

Grond L, Schulze P, Holstein J (2013) Systems analyses power-to-gas: a technology review. Groningen. http://www.dnv.com/binaries/dnv%20kema%20(2013)%20%20systems%20analyses%20power%20to%20gas%20%20technology%20review_tcm4-567461.pdf. Accessed on 30 Jan 2014

Karlen et al (2014) multilocation corn stover harvest effects on crop yields and nutrient removal. BioEnergy Res 7:528–539

Karp A, Richter GM (2011) Meeting the challenge of food and energy security. J Exp Bot 62(10):3263–3271

Koh LP, Ghazoul J (2008) Biofuels, biodiversity and people: Understanding the conflicts and finding opportunities. Biol Conversation 141(10):2450–2460

Krutzler Th et al (2015) Energiewirtschaftliche Szenarien in Hinblick auf die Klimaziele 2030 und 2050. Synthesebericht 2015. Umweltbundesamt

Nonhebel S (2007) Energy from agricultural residues and consequences for land requirements for food production. Agr Syst 94(2):586–592

Österreichische Energeiagentur (2015) Szenarien für die Strom- und Fernwärmeaufbringung 2030–2050 im Hinblick auf Klimaziele 2030 und 2050, https://www.energyagency.at/fileadmin/dam/pdf/publikationen/berichteBroschueren/Szenarien_fuer_Strom-_und_Fernwaermeaufbringung_und_Stromnachfrage_im_Hinblick_auf_Klimaziele_2030_und_2050.pdf

Pladerer Ch et al (2016) Lagebericht zu Lebensmittelabfällen und -verlusten in Österreich, Wien März 2016

Rathmann R et al (2010) Land use competition for production of food and liquid biofuels: an analysis of the arguments in the current debate. Renew Energy 35(1):14–22

Reiter G, Lindorfer J (2015) Global warming potential of hydrogen and methane production from renewable electricity via power-to-gas technology. Int J LCA 20:477–489

Reisinger H et al (2012) Rückstände aus der Nahrungs- und Genussmittelproduktion. Materialien zur Abfallwirtschaft. Umweltbundesamt, Wien 2012

Scherhaufer S et al (2016) Ökologische Auswirkungen von Lebensmittelabfällen

Statistik Austria (2017) Useful energy analysis. https://www.statistik.at/web_en/statistics/EnergyEnvironmentInnovationMobility/energy_environment/energy/useful_energy_analysis/index.html (dl: 07 June 2018)

Tilman D et al (2009) Beneficial biofuels—the food, energy, and environment trilemma. Policyforum 325:270–271

Vetter A et al (2012) Welchen Beitrag kann Biomasse zur nachhaltigen Bioenergienutzung leisten? Technologische Herausforderungen. Thüringer Landesanstalt für Landwirtschaft. https://www.energetische-biomassenutzung.de/fileadmin/user_upload/Intern/4._statuskonferenz/11051112_TLL_Statuskonf_Berlin.pdf (dl: 13 Feb 2017)

Zethner und Süßenbacher (2012) Vergärung von Wirtschaftsdüngern in Biogasanlagen. Evaluierung hinsichtlich Klimaschutzrelevanz, Umweltbundesamt, Wien, p 2012

Chapter 10
Manure Treatment for Biofuel

Willem van Laarhoven

Animal excrements, such as manure, are by far the largest organic waste stream in the European Union. It is particularly large in the Netherlands where millions of animals per year are raised for meat, milk, eggs, gel, leather and many other products for daily consumption. As long as people consume such animal products, this production causes an unavoidable (wet) stream of organic matter containing various minerals. The question is, whether the material content of manure can be recycled, meaning all its main components reused as valuable products. In this chapter, we address the possibilities of separating and reusing these components. The focus is on cattle and pigs because most of the dry manure of chickens is already largely reused (mainly for energy production), and the manure volume of other animals is insignificant.

10.1 Manure Supply

When farmers produce fodder for animals on their own land, they use manure to keep the land healthy and productive. In the latter part of the twentieth century, however, more fodder was and is being produced on an industrial scale outside farms, and so farmers need less land to produce fodder for their livestock. Since pig and chicken farmers hardly need land for fodder because they purchase it from the industries, the number of animals per hectare can multiply; this is observed, for example, in the sandy soil regions of the southern and eastern Netherlands. As the livestock per hectare increases, more manure is produced, which is difficult to dispose on one's own land, yet has to be processed or disposed of on other farms.

Manure is useful on farms. Intensive agriculture needs organic matter and minerals which are found in manure and mineral fertilizers. Excessive use of the mineral fertilizers, however, causes low carbon to nitrogen quotient (C/N quotient), which

W. van Laarhoven (✉)
Sint-Oedenrode, The Netherlands
e-mail: wvl@valacon.nl

Y. Krozer and M. Narodoslawsky (eds.), *Economics of Bioresources*,
https://doi.org/10.1007/978-3-030-14618-4_10

Table 10.1 Production of manure (× 1 million tons) and minerals in the manure (× 1 million kg) by dairy cattle, pigs and poultry in the Netherlands in 2016

	Manure	N	P_2O_5	K_2O
Cattle	61	307	93	414
Pigs	10.2	97	39	65
Poultry	1.4	62	29	29
Others	5.5	38	14	36
Total production	78.1	504	175	544

enhances the breakdown of the organic matter in the soil and therefore reduces soil fertility. This impact can be tempered by more use of manure, but too much can damage the plants. Maize, for instance, is tolerant to large amounts of manure but grains, potatoes, sugar beets and other plants grown on arable land become vulnerable with excess manure. Hence, farmers must balance organic matter and minerals. A surplus of manure, herewith, must be transported. This is costly especially in the case of wet manure; for instance, the cost of 150 km pig and cattle manure is similar to 300 km manure solids and 600 km dried manure pellets. These distances relate to the maximum of that which is feasible. Table 10.1 represents the production of manure and minerals by dairy cattle, pigs and poultry in the Netherlands in 2016. Nitrogen in the manure covers about 60% of the total input on arable land, and phosphate in the manure about 90% of the total.

The high use of mineral fertilizers and fertilizer in the manure caused severe environmental problems, such as saturation of land with phosphate, copper, cadmium and other heavy metals; pollution of groundwater and surface waters with phosphates, potassium and nitrates; and air pollution by NH_3, N_2O, CO_2, CH_4, odours and other pollutants that pose a risk to health and ecosystems. As plants require a balance between various minerals, the restriction of one mineral has implications for others. Regulations in the Netherlands restrict the use of phosphate minerals, which implies a maximization of the other minerals in manure. Nevertheless, nitrates are also maximized to resolve environmental problems in recent times (fosfaat—en stikstofgebruiksnormen, RVO.nl). In the last decade, losses to the environment have been decreased.

Ever since the 1990s, livestock farmers supported by related industries made attempts to find a manure treatment which could reduce transport, improve their use on arable farms (at home and abroad) as well as create markets for the disposal of manure. With the processing of manure, they expected to develop products that could replace mineral fertilizers and thus contribute to the resolution of some environmental problems; as the production of fertilizer is dependent on a lot of energy, causes pollution and involves exhaustible products such as natural phosphates. The digesting of manure would produce energy that could be used for one's own production and sales, thereby contributing to the reduction of greenhouse gases. Rigorous research and development was conducted, but lead to unsatisfactory results mainly because the costs of the processes exceed the income that could be generated from the sale of energy and minerals.

10.2 Energy from Manure

The primary purpose of processing manure is energy. Manure is digested, followed by the separation and filtration of organic matter, nitrogen, phosphate and other compounds. Digestion is a biological process which breaks down organic matter to deliver biogas. Biogas consists of 30–80% methane, 35% CO_2 and some 0–2% each of H_2S, NH_3 and H_2O; the methane percentage varies within the same type of manure and between types of manure. Biogas can be used as a fuel in a combined heat and power plant (CHP) for electric and thermal energy. Subsequent steps in the process are briefly described.

Biogas

Since digestion is a complex bacterial process, fluctuations must be managed so as to achieve a high output of biogas. The wet residue of digestion, called digestate, can be used on land as organic matter. Minerals in the digestate can be separated and used as replacements for mineral fertilizers. The energy of biogas can be used for that separation or it can be sold as heat and power. Table 10.2 indicates the methane production from cattle and pig slurry, calculated as the percentage of dry matter and organic matter in the dry matter. The dry and organic matters differ between these sources of manure, implying different outputs for biogas and methane.

Methane from cattle and pig slurry is small in comparison with other bioresources because their percentage of dry matter is relatively low; and bacteria degrade manure in stock. The longer the manure remains in the pit, the lower the methane production. Within two months, the biogas production amounts to half as much as fresh manure. Figure 10.1 shows the decrease in biogas for pig slurry. The economy of digesting depends on that lead time, implying that digesters close to the manure are most cost-effective.

Co-products

Biogas production only with manure (mono digestion) is uneconomic because organic matter is low and water content high (Kasper and Peters 2012). Co-products with higher organic matter are combined with manure; the potential energy generated by co-products can be up to forty times higher than manure (Zwart et al. 2006). The co-products are organic residues and plants. Organic residues of the food and beverage industries are available (van Oosterhout and van Laarhoven 2004; Vis 2002) as

Table 10.2 Typical percentage of dry and organic matter of cattle and pig slurries with biogas and methane production (Zwart et al. 2006)

	DM%	% OM	m^3 Biogas/ton OM	Biogas/m^3 manure	CH_4	M^3 CH4
Cattle slurry	9.0	6.6	300	19.7	60	11.8
Pig slurry	11.0	8.3	350	28.9	60	17.3

DM dry matter, *OM* organic matter

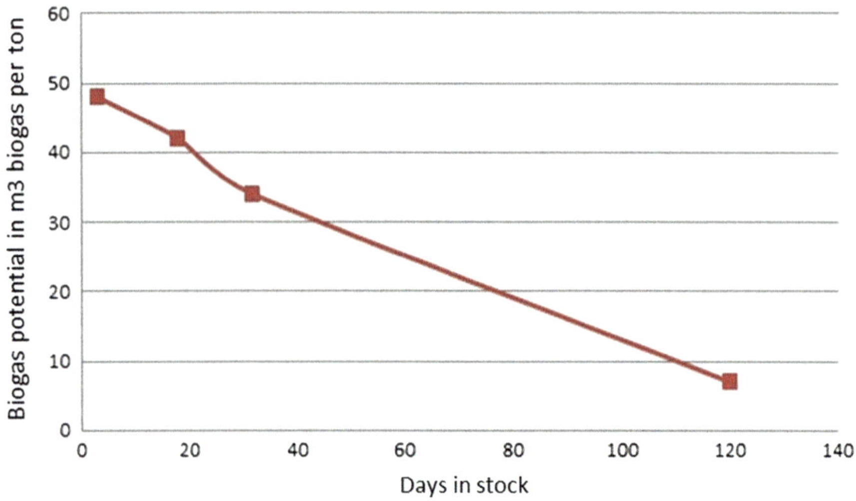

Fig. 10.1 Biogas potential of pig slurry, as a function of days in stock (De Buisonjé and Verheijen 2014)

Table 10.3 Some co-products, manure types and their potential for biogas and methane production (Veldhuis and van Laarhoven 2006; Wikipedia 2018)

Product (Mass)	Biogas in m^3 per ton	% CH_4	m^3 CH_4
Maize silage	202	52	105
Grass silage	172	54	92
Rye-GPS	163	52	84
Fodder beets	111	51	56
Bio residuals	100	61	61
Starch residuals	132	65	85
Bleaching clay residuals	455	60	273
Sewage sludge	75	57	43
Sugar beets residuals	67	72	48
Brewers grains residuals	40	61	24

they cover nearly 50% of all industrial waste in the Netherlands, measured by mass; of which 80–90% is used as animal feed, a co-product in digesters and for compost (Welink 2015). Various plants can be used as co-products. Table 10.3 shows the biogas and methane potential of some co-products.

Methane depends on the organic matter and degree of digestion. The addition of co-products delivers a nearly linear increase in energy; less slurry means higher energy production. The disadvantage of using co-products is that they contain minerals N, P and K as well as some undesired components which are restricted by regulation for reuse on land. The use of digestate with co-products, therefore, necessitates more land or less use of the mineral fertilizers. The latter reduces costs but entails effort from farmers in balancing nutrients.

Energy Production

The produced biogas can be converted to energy using a combined heat and power system (CHP), which consists of a gas turbine and electricity generator with a dynamo. CHP integrates the production of heat and electricity (power) in one single rather efficient process, because it captures high calorific heat for the generator. The thermal energy is about twice as high as the electric energy; the electric efficiency is 30–40% and that of thermal about 40%. The use of thermal energy at full capacity, minimum 7500 h a year, is key for the economic use of CHP. A maximum utilization is needed to cover high investment costs. A fluctuating production of methane from the digester, however, decreases the CHP energy efficiency. Thermal energy is also used for the drying of manure and co-products, as well as for management of the digestion temperature. It can also be used for killing harmful organisms termed the 'hygienization' of the digestate. While this is needed for the export of digestate, it is only a regulatory requirement that does nothing to improve the composition of digestate.

10.3 Processing of Digestate

Digestion hardly reduces the costs of transport because the volume of water in manure remains at about 90% of the slurry. Volume even increases when the co-product is added to the manure. That volume can be reduced by the mechanical separation of digestate into solid and liquid fractions. The solid fraction contains most of the dry matter and components bound to it; such as phosphates and organic nitrogen, as well as some potassium, but 50–80% of nitrogen and salts are found in the liquid fraction. If fresh manure is used, phosphates are less bound to the organic matter, and postseparation, they can be extracted from the liquid fraction by dephosphating.

Separation

Mechanical separation can be achieved by applying pressure with a screw press, belt filter press or centrifugal force (such as a decanter centrifuge, or a cyclone). The performance depends on the structure of the digestate, diameter of the filter and pressure applied. The separation efficiency measured as separated solids in the solid fraction varies for dry matter from 20 to 80% for phosphates; 35–90% for nitrogen; 25–50% for potassium from 15 to 25%. A higher efficiency delivers more dry matter and minerals into the solid fraction. In other words, more water with less dry matter and minerals are delivered.

The decanter, belt filter press and cyclones have the highest efficiency, but are more expensive than the screw press and need more energy. When aiming at an efficiency increase and more soluble components in the solid fraction for a cleaner liquid fraction, costly flocculants and coagulants are used as binders. The choice of technology used would depend on the requirements of the final fractions; will the liquid fraction be cleaned to the maximum possible extent for reuse, discharged in sewage or in surface waters? In answering this, let us examine a case study based on a decanter centrifuge.

Solid Fraction

Pelletizing

After the separation that delivers 30–80% dry matter, depending on the separation technique used, the solid fraction can be dried using thermal energy of the CHP and pelletized to 90% dry matter for use as fertilizer, then packed, stored and preserved in big bags. This can mitigate fluctuations in market prices caused by imbalances in supplies and demands. Pellets also meet the hygiene requirements for export.

Incinerating

The solid fraction has sufficient calorific value for co-incinerating with other fuels. Experiments evidence that possibility, however, the feasibility depends on the specification of dry matter because instabilities during the process of incineration can cause air emissions that contain chlorides, dioxins, NOx and ultrafine dust; all of which are pollutants. Solutions for those are costly (van Haaster and van Laarhoven 2004; van der Waall and Verberne 2004). Until present times, there was no initiative in practice on account of a rather low calorific value of the pellets.

Composting

In combination with composting, dry matter can increase to 35–60%. Herewith, hygienization is needed to meet the regulations for export. If one wants to export compost, it must be hygienized. Composting itself is an easy way to further dry the solid fraction. During composting, the temperature increases to 70 °C. It is also the hygienization of dry matter; if a higher temperature is needed, heat from the CHP can be used. While the volume does not change much, it matches export demand. This process requires little investment and energy, but rarely adds quality or market value.

Liquid Fraction

Dephosphating

Dephosphating refers to the chemical or biological extraction of phosphates. In biological processes, phosphate is consumed by bacteria under anaerobic conditions to produce polyphosphate. Bacteria and the polyphosphate sink and can be discharged by precipitation. Dephosphating by chemical treatment produces an insoluble precipitate of aluminium, calcium and iron together with phosphate, all of which is collected in a sink and discharged. Another technique is dephosphating with magnesium and ammonium that results in struvite ($MgNH_4PO_4 \bullet 6H_2O$) which contains

5.7% N, 28.9% P_2O_5 and 16.0% MgO. Struvite is a replacement for and is usable as mineral phosphate fertilizers (Motta et al. 2017). Despite the excess of phosphates in the soil and many secondary sources of phosphate, 27,000 tons of mineral phosphate fertilizers were used in the Netherlands in 2017 (www.kennisakker.nl, 2018). Other minerals remain in the effluent and can be used as liquid fertilizers, for example, potassium and nitrate. Morcatelli et al. (2017) concluded that digestion, separation and microfiltration can generate concentrates of minerals that are useful for drip irrigation.

Stripping Ammonia with CO_2

Nitrogen in manure consists of about 50% water soluble ammonia, and 50% nitrate bound to the organic matter. Most of the latter is separated from the organic matter early in the separation process. The soluble nitrate remains in the effluent. It can be separated by stripping and reused as a fertilizer that replaces mineral fertilizers. This technique uses chemicals such as caustic soda and sulphuric acid. Ammonia is driven into the gaseous phase (stripping) with the caustic soda and then converted (scrubbed) with sulphuric acid to get ammonium sulphate. This way, about 65% to 80% of the dissolved nitrogen can be discharged. The efficiency can be increased by using catalytic oxidation with platinum, under higher temperatures which uses about 20 kWh per ton of effluent. Research shows that stripping with acid can result in a lower availability of nitrogen for plants in a short growing cycle, and that it can cause leaching into the groundwater (Sigurnjak et al. 2017). An alternative is stripping with CO_2 for ammonium bicarbonate ($2NH_3 + CO_2 + H_2O \rightarrow (NH_4)_2(CO_3)$). This can be split into ammonium and carbonate. Using nitrification, the ammonia can be converted to nitrate and used as a liquid fertilizer. When CO_2 is used as a regenerative carrier of ammonia, no additional chemicals are needed. Alternatives for nitrogen reuse are researched (Manure resource 2017).

Nitrification and Denitrification

Nitrification and denitrification are biological processes that can be used to remove nitrogen (nitrate and ammonia) from an effluent. The first step is aerobic nitrification of ammonia into nitrite (NO_2), and then into nitrate (NO_3). The process requires a lot of oxygen, brought into the effluent as a result of high energy use. Then, in the anaerobic step, nitrate is converted to gaseous nitrogen (N_2) that dissolves in the atmosphere. A relatively new process is Anammox (ANaerobic AMMonia OXydation). It removes nitrogen without aeration at a temperature above 25 °C, using thermal energy from the digester. Its energy consumption is comparatively lower than that of aerobic nitrification, but the concentration of ammonia must be high; costs are low due to compact installation. So far, this technique is not used for the treatment of effluents from the digestate but can be combined with other nitrification processes.

Filtration

After the discharge of 50% or more of the organic matter, nitrogen and phosphate from the liquid fraction of the effluent still contain components such as minerals, salts, humic acids and organic residues. A clean effluent that can be reused or discharged into the surface water needs filtration. There are different techniques for various sizes of particles in the effluent; in ascending order of filtration, they are microfiltration (MF), ultrafiltration (UF), nanofiltration (NF) and reverse osmosis (RO). Smaller openings in the membranes need a higher pressure for separation of smaller particles, which require more energy.

There are dead-end and cross-flow techniques. With dead-end filtration, all the water runs perpendicular to the membrane surface and the separated particles build up, over time, on the surface of the membrane (retentate). When the particle layer thickens, the membrane has to be cleaned or replaced until the next batch. With the cross-flow filtration technique, the water flows tangentially across the surface of the membrane leaving no retentate, as it is washed away during the filtration process. There is no need to stop the process for cleaning; it is continuous and used for liquids with very small particles. Microfiltration is cheap, reliable and used for liquid fraction of the decanter. Reverse osmosis operates on the cross-flow technique, and it is costly but useful to clean effluents with small particles; effluents of microfiltration are desalinated.

Evaporation

The concentrated matter that remains after these filtration steps still contains water. The water content impedes the replacement of mineral fertilizers. Evaporation can reduce water to less than 20% moisture. Mechanical vapour repression (MVR) and thermal vapour repression (TVR) are the most popular techniques. Both use thermal energy of the CHP for drying, up to about 80% dry matter. However, the techniques are expensive and energy-intensive. They use about 2.3 GJ per ton evaporated water, though the efficient techniques in the dairy industry use about 1.0 GJ per ton evaporated water. When MVR and TVR techniques are combined, the energy use declines to 0.27 GJ per ton evaporated water. The higher the transport costs of end products, the higher is the profitability from evaporation.

10.4 Ecological Aspects

Impacts should be assessed in order to answer the question of whether manure treatment is environmentally friendly and can save fossil energy. Manure treatment has various direct and indirect environmental impacts at different stages of its handling and storage. The most important of these are:

- The duration of manure storage and therewith the emission of ammonia, nitrogenous gas and methane.

- The production and use of co-products and therewith the use of energy, fertilizers and soil born emissions.
- The transportation of manure and co-products and therewith the use of fossil fuels and CO_2 emissions.
- The replacement of fossil fuels by using renewable energy from biogas and therewith the emission of CO_2.
- The effects of using end products of the processed manure instead of mineral fertilizers and therewith the energy consumption and use of resources (as the production of N-fertilizers needs much energy, and the production of P-fertilizers based on the phosphates mining is polluting and leads to scarcity).

Untreated Manure

Low organic matter on arable land, even in regions where a lot of manure is used is a major concern of the agricultural sector and researchers (Conijn and Lesschen 2015). Degradation of the organic matter on farmland causes a loss of fertility of the land, degraded structures of water and minerals capture as well as lower crop yield. Loss of minerals, CO_2, gaseous nitrate and methane can be expected, which adds to emissions of greenhouse gasses. More organic matter enhances the biomass production due to better fertilization (short C-cycle), which contributes to the capture of CO_2 as well as other environmental benefits. The overall impact depends on the possibilities of replacing mineral fertilizers with manure, which relates to the type of crop, season and weather during manure application.

Digestion with Co-products

Digestion of cattle and pig manure alone is neither ecological nor economic (Zwart et al. 2006; Vos and Zwart 2013; Rabo 2012a, b) because the energy content of slurry is low compared to the energy required for the slurry transport and processing. A higher energy efficiency is attained with co-products of plants such as rye, barley and maize or with organic residues of the food and beverage industries. The three options are:

1. Co-digesting manure with specially grown products
2. Co-digesting manure with organic residues
3. Co-digesting manure with a combination of the above.

Digesting with Plants

If it is assumed that manure is digested fresh without the need for storage, then only a small part of its energy content is used for the process and the surplus can be sold. Biogas surplus replaces fossil energy, which reduces CO_2 emissions. Digestion of slurry alone, however, is not feasible because of low organic matter in manure. The ratio of manure to co-products determines the energy efficiency and environmental impacts of manure.

Digestion of only manure and use of this digestate for cultivation reduces CO_2 emissions by 90%, as compared to non-digested manure; this number increases when

Table 10.4 Emissions of greenhouse gases when slurry is digested with maize as a co-product on a 50/50 basis, compared to the undigested manure; the input for the digester is based on a CHP capacity of 0.5 MWh (Zwart et al. 2006)

Combinations	Manure (tons)	Maize (tons)	CH_4 (tons)	N_2O (kg)	CO_2 (tons)	NH_3 (kg)
Non-digested pig slurry	42,500		143.3	332.2	3112	6664
Digested pig slurry	42,500		10.4	16.6	223	339
Pig slurry and maize digested	4266	4266	4.1	406.8	349	1362
Non-digested cattle slurry	36,360		122.6	158.7	2624	3174
Cattle slurry digested	36,360		9.4	7.9	199	190
Cattle slurry and maize digested	4195	4195	13.4	416.7	344	1670

methane in manure is effectively converted into biogas. However, positive effects decline if co-products are used. When plants as co-products are involved, minerals are used during cultivation, storage and transport which cause the CO_2 emissions. If the digestate volume is not decreased, its transport is energy intensive, and an extent of the minerals in the digestate is lost to the environment when used as fertilizer. More plants as co-products cause additional emissions because of the plants' cultivation. Table 10.4 shows environmental impacts.

In general, when more manure is used for digestion, lower is its negative environmental impact; except for NH_3 because greater co-products reduce its emission. Also, more CO_2 can be expected when many co-products are used, even though fossil fuels are substituted for methane in biogas. The more co-products that are used, fewer are the environmental advantages.

Digesting with Residues of Food and Beverage Industries

Residues of the food and beverage industries deliver more methane than manure. When compared to the plants used for co-products, the use of organic residues cause less impact if crops for the fodder are grown; crops such as maize or rye which adds emissions. Given that the production of fodder also needs energy, digesting with organic residues reduces the positive energy effects of digestion. In addition, the monetary value of fodder is higher—than methane from digestion in biogas—which implies that the cultivation of fodder for digestion is uneconomic.

Environmental Impacts of Co-digestion

When the production of co-products is taken into account, the environmental impacts depend on the: (a) ratio of manure to co-products and ways in which these are produced and stored (as this determines the methane production); (b) energy-efficiency of the co-generator for power and heat; (c) transport of manure and co-products; and, (d) substitution of the mineral fertilizers with the products of manure processing. The last point is important because weather, soil conditions, and the type of crop influences environmental impacts.

10.5 Economic Impacts

Several studies show that digestion, with or without co-products and with or without separation of minerals, is unfeasible (Rabo 2011; Vos and Zwart 2013). The digestion could be feasible if the disposal costs of manure are included; which in the Netherlands were € 18–€ 20 per ton (van der Voort et al. 2008). Other studies suggest that the right mix of co-products and efficient CHP can be cost-effective, but the scale of processes and co-products must be large (Zwart et al. 2006). Herewith it is pinpointed that the availability of co-products is limited to 10% of all manure, and feasibility of using them for digestion depends on prices and subsidies (Peeters et al. 2012). The use of crops for composting and fodder pushes the prices of co-products up. Furthermore, low profitability of grain and maize for co-digestion compared to other crops reduces the farmers income (van der Voort 2007), whereas high prices lead to unfeasible manure digestion (van der Voort et al. 2008). Fluctuating quality and composition of the co-products also cause laborious CHP operations, even its contamination; and the fluctuating flow of biogas to the CHP causes its underutilization (Veldhuis and van Laarhoven 2006). All factors considered, the high costs of co-products determine the economics the manure digestion and is considered unfeasible. Appendix 10.1 shows exemplary calculations.

Optimization

Mono digestion is unfeasible because when undertaken in the large scale, it neither increases income, nor is downscaling a remedy. An improvement to be made could be a higher production of biogas due to a shorter duration of residence of substrates in the digester.

The addition of glycerine fosters digestion, but is costly and causes imperfections in the digestion processes (Kasper and Peters 2012; Kasper and Kemperman 2015). For instance, 4–5% glycerine can shorten a 60 days digestion period by 10–14 days, as well as increase the output; but the price remains high while the quality of biogas fluctuates (Timmerman et al. 2013).

Another adaptation could be adding a pre-digestion step to the process, which facilitates the breakdown of organic matter to 70–75% as compared to the usual 50%, thereby shortening the digestion time and increasing the capacity. This is called a 'two-phase system' (Zwart et al. 2006). In the first phase of digestion, the hydrolysis and production of volatile fatty acids is intensified in a separate step in the process. This way, the capacity of the digester can be increased 8–15 times, and therewith also the energy production. However, this would require a modified process layout which is not feasible on the farm scale. The affordability of digesting can—nonetheless—be increased.

Another optimization option is to bring back the digestate into the digester for a 'second chance'. This action enables the breakdown of the organic matter by 60%. Alternatively, digesting of the solid fraction with 35% dry matter instead of the usual 9% after separation can increase the capacity and energy production. It can be combined with liquid organic co-products from industries. While a higher temperature intensifies the process, it uses more thermal energy.

The key factors in digestion for biofuel are tuning the availability of co-products with the scale of CHP for full scale operations during more than 7500 h per year. The digestion can generate income only with an efficient utilization of the thermal and electrical energy of the CHP; each one covers about 50% of the total energy production. Use of solely thermal energy does not generate sufficient income because heat losses in transport are incurred. Heat can be used for drying and steam pressing of pellets or other products from the digestate, but the use of heat in the processes is less economic than the sale of heat.

Manure Chain

The cost of co-products and sale of energy and end products are key variables that determine the feasibility of manure recycling (Smit et al. 2012). Hence, only a cooperative approach in the chain—right from the purchase of co-products down to the sale of energy—can deliver profits. The profit can be divided among participants in the chain in relation to their contribution to it.

A few options for an illustrative biogas installation are assessed with the aim to show sensitivity for those key variables already mentioned. The following assumptions are made—the installation has an average capacity of 36,000 tons a year. As the disposal of manure is a cost to farmers, a rate of € 15 per ton for manure delivery is assumed. This is the usual price farmers pay for excess manure in the Netherlands; a price any lower than this would imply unfeasible manure recycling. The estimate is based on the digestion of 18,000 tons manure and 18,000 tons co-products from residues of the food and beverage industry. Two purchase prices of those co-products are considered: € 120 and € 97 per ton residue. The price of the manure replacer is assumed equal to the mineral fertilizers. Results are shown in Tables 10.5 and 10.6. It shows an estimate of a situation in which the processing of manure and co-products is imperfect, called 'sub-optimal'. Another estimate is based on hypothetical ideal processing called 'optimal', which though ideal is rarely found in practice. The third estimate approaches the processing that is often experienced in practice, called 'middle'.

The assessment reveals that manure recycling with co-products is feasible only under optimal conditions. Moreover, the processing is cost-effective only when the

Table 10.5 Estimates of manure processing (based on Zwart et al. 2006; Veldhuis and van Laarhoven 2006)

	Sub-optimal	Optimal	Middle
Biogas production			
Cattle slurry (m^3 biogas/ton)	200	400	300
Pig slurry (m^3 biogas/ton)	200	500	375
Maize (silage) (m3 biogas/ton)	400	650	575
Reye (silage) (m^3 biogas/ton)	300	600	450
Barley (silage) (m^3 biogas/ton)	230	1100	650
Residual from food and beverage industries (m^3 biogas/ton)	400	1.200	800
Dry matter of the co-substrate in %	10%	30%	25%
Organic matter % of the dry matter	70%	85%	80%
Methane in % of the biogas (%)	47%	70%	60%
Process parameters			
Running hours CHP (h/year)	5.000	8.100	7000
Total energy conversion CHP	60%	75%	70%
Electric efficiency CHP	35%	42%	40%
Thermal efficiency CHP	25%	33%	30%
Internal use electric energy (MJ/ton)	35	30	32
Internal use thermal energy (MJ/ton)	275	225	250
Utilization of thermal energy (%)	50%	90%	80%
Energy leakage installation (%)	5%	1%	2%
Economical parameters			
Gate fee manure	€ 0.00	€ 15.00	€ 15
Acquisition costs co-products (€/ton)	€ 120.00	€ 50.00	€ 120
Price electric energy (€/kWh)	€ 0.05	€ 0.05	€ 0.045
Price thermal energy (€/KWh)	€ 0.05	€ 0.06	€ 0.055
Internal energy price (€/kWh)	€ 0.030	€ 0.025	€ 0.27
Costs of pelletizing (€/ton)	€ 16.00	€ 12.00	€ 14.00
Price of fertilizer pellets (€/ton)	€ 80.00	€ 150.00	€ 100
Costs for disposal digestate (€/ton)	€ 30	€ 10	€ 20

Table 10.6 Assessment results 15/97 and 15/120 are fees for the manure € 15 per ton and for co-products, € 97 per ton and € 120 per ton, respectively

All in €	Optimal 15/97	Middle 15/97	Sub-optimal 15/97	Sub-optimal 15/120	Middle 15/120
Benefit					
Gate fee manure	270,000	270,000	270,000	270,000	270,000
Energy	1,130,216	235,100	−202,483	−202,483	235,100
Fertilizers	1,121,976	925,976	741,976	741,976	925,976
Total	2,522,192	1,431,076	809,493	809,493	1,431,076
Benefit per ton input	70.06	39.75	22.49	22.49	39.75
Costs					
Purchase co-product	1,746,000	1,746,000	1,746,000	2,160,000	2,160,000
Operation digester	318,240	330,630	347,490	347,490	330,630
Operations total	424,650	438,634	402,757	402,757	438,634
Discharge effluents	32,864	41,080	42,400	42,400	41,080
Total costs	2,521,754	2,556,344	2,538,647	2,952,647	2,970,344
Costs per ton input	70.05	71.01	70.52	82.02	82.51
Net result					
Income €/ton	0.01	−31.26	−48.03	−59.53	−42.76

prices of co-products do not exceed € 97 per ton. A sensitivity analysis shows that even if the co-product is free of charge, the process must be optimal to generate a net positive result of € 24 per ton. These results are in line with other studies (Zwart et al. 2006; Peeters et al. 2012; Vos and Zwart 2013).

Feasibility of manure recycling also depends on the political setting. A lot has been studied, but vested interests in policies, businesses and environmental organizations reinforce positions entailing regulations that rarely contribute to a sustainable solution. If the agricultural businesses want to resolve its problem of excessive manure production they should enhance economy of scale, optimal technologies and marketing of the end products (Nest Vanden 2017). There must be a will to cooperate, even with opposing opinions.

10.6 Conclusions

Cost-effective processing of manure is possible but only with a cooperative approach throughout the value chain of suppliers, and sales of energy. Such a cooperative approach also generates environmental benefits. If the benefits are awarded, for instance due to liabilities for pollution, the economic and environmental interests for the recycling of manure increase and the manure processing becomes feasible. A price on CO_2 emissions, even as low as € 15 per ton, would make several options feasible as energy sales would be more attractive. The cooperative approach in the chain is possible when key parties show their commitment, and there is a marked moving away from conventional thinking. This distinct change should be concerned with policy and a shift of focus within agricultural organisations, from own profits to profits in the chain. It is possible because agricultural business hails from a tradition of co-operation. Experiments in the regional chain are recommended as well.

Appendix

Tables show options of manure processing and the typical cost of a CHP.

Economic parameters of (co-)digestion of manure and maize silage in different scenarios for the Dutch province of Overijssel (Vos and Zwart 2013)

Scenario	Biomass	Investment (€)	Specific investment (€/kW)	Profit per year (€)	Ratio cost to income
1,7 MWe	32,000 tons of ECN Mix[a]	4,712,055	2772	−922	2.2
Low price biomass	180,000 tons of fresh pig slurry	9,994,970	5879	−1,844,073	3.7
Scaling down (CHP)	5000 tons of fresh cattle manure	255,444	6904	−37,801	2.3
Scaling down (green gas)	5000 tons of fresh cattle manure	438,849	11,861	−59,303	4.4
Scaling up	100,000 tons of ECN mix[a]	15,961,652	3192	−3,036,324	2.2
Extra delivery of thermal energy (factor > 3)	32,000 tons of ECN mix[a]	4,821,714	2836	−761,632	1.8

[a]ECN mix: 25% manure from own farm + 25% manure from elsewhere + 13% silage maize + 37% other co-substrate

Typical technical and financial parameters of a CHP-production in 2010; unit costs in eurocents per kWh (Rabo 2011)

Technical parameters	2010
CHP capacity (MWe)	1.5
Investments/MWe	2,900,000
Total running hours CHP for full for load per year	7300
kWh production/MWe CHP	7,300,000
% Energy use by the process	8.0
Net energy production kWh/MWe CHP	6,700,000
Economical parameters	
Other incomes[a]	−0.2
Costs for co-products (Input + Output)	7.8
Maintenance costs	1.8
Labour/consultation/others	1.8
Depreciation and interest	5.3
Total costs	16.5
Cost price/kWh delivered on the energy grid	18

[a]Not including energy sales and subsidies (mostly thermal energy)

References

Conijn JG, Lesschen JP (2015) Soil organic matter in the Netherlands. Quantification of stocks and flows in the top soil. Plant Research International/Alterra, WUR Wageningen

De Buisonjé F, Verheijen R (2014) Drijfmest verliest snel zijn waarde voor biogas. Wageningen UR Livestock Research, Varkens Innovatie Centrum Sterksel. V-focus April 2014, pp 20–21

Kasper GJ, Kemperman A (2015) Mestvergisting Fermtech Systems en Ecobag. Rapport 874 Wageningen UR Livestock Research

Kasper GJ, Peters B (2012) Monovergisting varkensmest op boerderijschaal. Rapport 632 Wageningen UR Livestock Research

Manure Resource (2017) 3rd International conference on manure management and valorization, November 2017, Eindhoven (NL)

Morcatelli G, Mantovi P, Piccini S (2017) A farm scale process based on digestate microfiltration aimed at fertigation through drip lines. In: 3rd International conference on manure management and valorization, November 2017, Eindhoven (NL)

Motta SR, Tambobe F, Adani F, Brenna S (2017) Phosphorous fertilizers efficiency compared in agronomic trials: struvite an opportunity of reuse and recycle. In: 3rd International conference on manure management and valorization, November 2017, Eindhoven (NL)

Nest Vanden T (2017) Tailor made digestate products by blending with animal manure products, compost and mineral fertilizers. In: 3rd International conference on manure management and valorization, November 2017, Eindhoven (NL)

Peeters SJW et al (2012) Achtergrondrapport integrale visie duurzame drijfmestverwaarding LTO. Energy Matters, Driebergen

Rabo (2011) Vergisting benchmark (co-)vergisting boekjaar 2010: rendement door markt in verdrukking! Rabobank Food & Agri Thema-update, Utrecht

Rabo (2012a) Benchmark (co-)vergisting boekjaar 2010: rendement door markt in de verdrukking. Food & Agri Thema update vergisting, 4e jaargang, November 2011. Utrecht

Rabo (2012b) Toekomst biogas: van laagwaardige input naar hoogwaardige output. Food& Agri Thema biogas update, 37e jaargang, januari 2013. Utrecht

Sigurnjak I, Michels E, Crappe S, Buysens S, Tack FMG, Neve S, Meers E (2017) Can acidification increase nitrogen fertilizer value of bio-based fertilizers? In: 3rd International conference on manure management and valorization, November 2017, Eindhoven (NL)

Smit A, Rulkens W, Sanders JPM, Verdoes N, Teng Y, Brunt D (2012) Verwerking van digestaat van Mestvergisting. Terreinverkenning van mogelijkheden tot kostenreductie. WUR/Alterra rapport 2310

Timmerman M, van Eekert MHA, van Riel JW, Schuman E (2013) Mestvergisting bij korte verblijftijden. Rapport 729 WUR Wageningen

van der Voort MPJ (2007) Saldoberekening van energiegewassen voor covergisting en bioethanolproductie. PPO Publicatienr. 32500580 Wageningen

van der Voort MPJ et al (2008) Economie van energiegewassen. PPO Publicatienr. 32500608, Wageningen

van der Waall WR, Verberne RA (2004) Multi-purpose pellet facility and real options portfolio management (ROPM) in response to value chain changes of biomass resources. Ingenia Consultants & Engineers, Eindhoven

van Haaster J, van Laarhoven W (red) (2004) Samengestelde biomassapellets. Coalitievorming voor grootschalige experimenten voor de meestook van samengestelde biomassapellets in energiecentrales. Rapport in het kader van de regeling BSE Ondersteuning Transitie-Coalities Duurzame energiehuishouding. Novem rapport 5005-03-20-01-013. FIS, Wageningen/Aarle-Rixtel

van Oosterhout T, van Laarhoven W (2004) Het gebruik van reststoffen van de voedings—en genotmiddelenindustrie in de veehouderij. Stuurgroep Technology Assessment ministerie LNV/Sirned. Den Haag/Ulvenhout

Veldhuis J, van Laarhoven W (2006) Nabewerking van digestaat na vergisting. Onderzoek naar de technische en financiële haalbaarheid van het naverwerken van het digestaat van de vergisting van natte restproducten uit de voedings- en genotmiddelenindustrie, Hengelo (OV)

Vis M (2002) Beschikbaarheid van reststromen uit de voedings—en genotmiddelenindustrie voor energieproductie. BTG/Novem, Enschede/Utrecht

Vos J, Zwart KB (2013) Mest(co)vergisting en biogas/groengas productie in Overijssel: Ervaringsproblemen, kansen & Verbeterstrategieën. BTG biomass technology group BV/Wageningen UR/Alterra rapport

Welink JH (2015) Meer waarde uit de reststromen. Toekomstverkenning van mogelijkheden recycling reststromen uit de voeding—en genotsmiddelenindustrie, TU Delft

Wikipedia (2018) https://nl.wikipedia.org/wiki/Biogas

Zwart KB, Oudendag DA, Ehlert PAI, Kuikman PJ (2006) Duurzame covergisting van dierlijke mest. WUR/Alterra rapport 1437, Wageningen

Part III
Reviews of the Sectors

This part presents three sectorial reviews on the production and uses of bioresources, with the aim to pinpoint ample opportunities for valuable activities when sustainable innovations are pursued. First, possibilities for the sustainable production of bioresources in agriculture are shown if sustainable innovations are embraced and supported by policies. Second, ample opportunities for the upgrade of low-value residuals to high-value products (in industries) are presented if R&D can deliver beneficial results because organic, recyclable materials are awarded on markets. The third review presents a few applications of bioresources in services, which underpins the growing consumer demand for natural resources, in particular bioresources which are found, if innovative efforts are focused on market niches.

Chapter 11
Innovating for Sustainable Agriculture

Willem van Laarhoven

Bioresources can be produced endlessly if nature is used with care. However, today's farming involves linear chains of producers in which animal and mineral fertilizers, pesticides, and other materials are supplied externally with much energy use and pollution. In addition, large material losses to the environment are observed and residuals are inefficiently reused. Possibilities to enhance the circularity in the agricultural chain are addressed because it is a key factor in the development of sustainable agricultural practices.

11.1 Impacts of Linear Agriculture

Agriculture is the main producer of bioresources, particularly food. With regard to the growth of the world population to 9 billion people, the demand for food will double in the next 50 years; and for that of meat and dairy will increase even more (Steinfeld and Gerber 2010). This implies the need for twice the agricultural production than the total agricultural production that was generated throughout the past 10,000 years (Eggen 2012). Presently, intensive agriculture involves large inputs while low material-efficiency depletes natural resources for instance phosphorus, and the production of N-fertilizers uses much fossil fuel and causes an environmental impact (Cordell and White 2014); livestock is particularly polluting. The global agricultural system is a major cause of climate change, depletion of clean water, contamination of water and soil and degradation of ecosystems with excessive use of nitrogen and phosphorus (Campbell et al. 2017). Given that the agricultural value chains are global, such impacts are also observed across countries, along all the

W. van Laarhoven (✉)
Sint-Oedenrode, The Netherlands
e-mail: wvl@valacon.nl

Y. Krozer and M. Narodoslawsky (eds.), *Economics of Bioresources*,
https://doi.org/10.1007/978-3-030-14618-4_11

chains (Foley et al. 2005). The impact is going to increase by 50 to 90% between 2010 and 2050 unless innovations aiming at circularity in agriculture are adopted (Springmann et al. 2018).

Large-scale monocultures are developed over the years, for instance, huge areas with only corn, soy, grains, or other crops. They deliver a large output (measured by biomass) but cause negative effects on continuity of agricultural production and on the environment. Equipment is enlarged and tuned to such monocultures that undermine soil structure as the heavier and more powerful machines that are in use compact the soil. Soil quality and the organic matter supplied to it are of concern. The water storage capacity has decreased, and in wetter times, the soils are difficult to process. Since soils can hold less water, there are shortages of water in dry times. Furthermore, fields have to be irrigated mechanically which in turn increases the extraction of water from the deeper groundwater and of surface water. In recent years, more attention has been paid to technical interventions in solving these problems, but they are overtaken by even larger machines. They emit even more greenhouse gases because the energy consumption is higher. Pests and diseases are aggressively spread, which prompts an increased use of chemical pesticides. The decrease in soil quality and fertility in monocultures leads to lower biodiversity and utilization of nutrients, which is compensated by more fertilizers or animal manure and pest control with chemical pesticides. Hence, the losses to the environment increase in all manners.

Fostering sustainable agriculture requires an integrated approach because changes in one step of the agriculture value chain affects others. Rather than adding more technologies to correct inefficiencies and suppress impacts (but they also complicate and increase the need for control in those value chains)—innovations are necessary for measures on the farm level and in the value chain, simultaneously (de Goede 2014). Policies aiming to reduce these impacts of agriculture are necessary; nonetheless regulations can enlarge costs and trigger the scaling up of firms, material use, and waste, while hardly any company turns to sustainable production. So, auxiliary activities and policies are also required. Economic incentives that forge new food patterns prevent product and energy losses, and foster closed production cycles and material efficiency are preferred so that the innovating companies can benefit. This way, knowledge about good practices disseminates, damages are priced (de Vries 2018) and innovative concepts in agriculture—such as vertical urban farms—stand a chance (Kalantari et al. 2017). Progress in sustainable agriculture involves better performance in production, co-operative practices in the value chain as well as policies that foster positive responses of consumers and citizens (de Wolf et al. 2018; UDV 2018). It can be monitored using a few key indicators as shown in Table 11.1.

11.2 From Material Efficiency to Innovative Systems

Individual farmers usually increase materials as long as extra income exceeds extra costs, which generate higher yields but lower income as per material costs. When many farmers do it, prices fall entailing losses, mergers, and ultimately bankruptcies.

Table 11.1 Selected key indicators for sustainable agriculture, based on Reytar et al. (2014)

	Policies	Practices	Performances
Water use	Policies[a] on abstraction	Area[c] efficient irrigation	Area drip irrigation or water use to stress ratio
Climate change	Policies on low GHG[b] agricultural	Share farms with GHG management	Foods production per GHG unit
Land conversion	Policies on ecosystem conversion to farms		Area conversion of ecosystems to farms
Soil health		Share arable land with conservation or share cropland	Area eroded, change of NPP[d] or organic matter
Nutrient pollution	Policies on soil conservation	Area with nutrient management	Nutrient balance or fertilizer use per area
Pesticides pollution	Policies on pest management	Share area with integrated pest management	Pesticide use per unit arable area

[a]In all cases existence of the policies
[b]GHG means greenhouse gasses
[c]In all cases share of the land
[d]NPP (Net Primary Production)

Despite losses in agriculture as a whole, individuals maximize their gains. Lack of co-operation in agriculture renders corrective regulations.

Farmers innovate when the costs of materials grow faster than that of revenues. Herewith, innovations in the farming systems refer to various types of farming aimed at balancing material inputs with harvests; particularly generating crops while maintaining organic matter for soil fertility, but with less mineral inputs and even without an external supply of nutrients. Such generations involve greater knowledge about the farming system and exert less impact on nature and environmental qualities; but usually produce less products per arable land though not necessarily lower income when higher product prices can be paid. Take for example organic dairy farms that are frontrunners in the recycling of materials from their own land, even though some external materials are needed on all farms.

Low-input Farming

In low-input agriculture, the aim is to limit the use of materials and auxiliary substances such as energy, fertilizers, animal feed, pesticides, and medicines to a necessary minimum. Also, the goal is to reduce production costs as well as environmental impact. The self-sufficiency (robustness) of the system would thus increase as would the profitability in the short and long term (Parr et al. 1990). It is not primarily about low external input rather about combinations that increase the internal input for improving soil fertility, better use of crop residues and animal manure, growing ground cover plants, and preventing the leaching of nutrients (Norman et al. 1997).

A higher material-efficiency, however, does not necessarily generate more income per animal or input. Comparisons of high-input companies with low-input ones within the framework of the European Sustainable Organic and Low Input Dairying (SOLID) shows that low-input companies are usually smaller and less productive because they deliver less products and their costs per unit labor, land, or capital are higher.[1] Some low-input companies are more productive but the factors that drive productivity and the differences between companies are difficult to establish (Leach 2012). Hence, policies aiming at low-input agriculture across all types of agricultural companies would be less productive, but a company-specific approach and measures could be more promising.[2]

Organic Agriculture

In the European Union, the area under organic agriculture and being used for the consumption of organic products (as measured by the certified areas and products) has increased. Organic agriculture does not allow inputs like that of mineral fertilizer, pesticides, and antibiotics. In addition, it aims to be as self-sufficient on the farm level as possible by optimizing the internal inputs for health of the soil and livestock; also by adjusting production output to the possibilities within the farm system. However, the outputs per area are usually lower, and more land is used for the cultivation of their own feeds. Hence, it is often debated whether this type of agriculture—if implemented on the global scale—can feed all people; though its superiority with respect to biodiversity on farms when compared to conventional agriculture is widely acknowledged. Taking away all mineral inputs from all farms would certainly cause a substantial drop in product outputs, but such drastic changes are not needed because the growth of organic farming can be focused on particular crops; for instance, organic olives are considered nearly as productive. In addition, the decline of outputs can partially be compensated through a decrease of agricultural and food waste; lower animal consumption per person; as well as technological changes in organic farming such as improved varieties, breeding practices, and integrated pest controls, which generates higher efficiency. Dairies often focus on the cultivation of crops as fodder to meet their own protein requirements. Higher sales prices of products enable the economic sustainability of organic agriculture. Integration of activities within a farm is essential to keep the cycle closed and is therefore also costly as it necessitates much knowledge about the system.

The sustainability of organic agriculture from environmental, animal welfare, and working conditions' perspectives is assessed; the result is that it is as better off due to less external medicines, energy, and nutrients on the farms. However, nutrients leach and ammonia is emitted because animals are likely to be kept outside. Low use of medication and a less conditioned stay increases the risk of diseases and disorders for some animals (Spruijt-Verkerke et al. 2004). The argument holds true

[1] http://www.solidairy.eu.

[2] https://www.ilvo.vlaanderen.be.

that organic farming cannot meet the growing food demands—given the present farming practices and certification rules—of the world population, because more land (that is less productive) is consumed per product; but better practices and wiser rules are possible.

Closing of Cycles

Intensive agriculture is inefficient with respect to carbon storage, recovery in organic matter, and the reuse of nutrients for plants, in particular nitrogen and phosphorus; but its efficiency can be improved by closing the carbon and nutrient cycles. The closing of cycles means reusing of the residual streams of vegetable crops from the food industries for animal feed, followed by the reuse of animal manure for soil improvement and as fertilizer for the cultivation of crops. Linking of crop cultivation and animal production implies that mixed agricultural farms can close the cycles easier than specialized ones. The optimization, however, depends on the quality and quantity of the residual streams and the needs of livestock farms, as well as the locations of the mixed production and the food-related regulations. An optimal balance of these factors would reduce the supply of raw materials from far away, and reduce losses to the environment, for instance the recovery of nitrogen and phosphorus would significantly improve (Lesschen 2018).

Recycling in agriculture, in particular that of nutrients, is often pinpointed from the environmental perspective with the objective of keeping the flows of nutrients and carbon within the farming systems, and to minimize the biomass losses within a farm or value chain (Scholten et al. 2018). This also reduces the dependency on materials from outside the food system, but needs co-operative efforts and the balancing of minerals and nutrients across sectors. Although circular agriculture is sometimes perceived as the ultimate form of sustainable agriculture (de Boer and van Ittersum 2018), closing cycles on the company level is either rarely possible or unnecessary because better nutrition, fertilization and good soil management reduces losses. The closing of cycles on a regional scale is more promising because different farms can jointly achieve better nutrient utilization, for example, due to co-operation between livestock farms and arable farms.

Optimizations within the dairy farm and the co-operation between a dairy farm and an arable farm are studied. Opportunities for better carbon and nitrogen balance are found on the dairy farm due to crop rotation, for instance, grasslands with maize and potatoes. Co-operation between a dairy and an arable farm can complete the crop rotation, which is also achievable on mixed farms that rotate the crops (de Wolf et al. 2018). However, such co-operation between livestock and arable farming is no guarantee for sustainability, because it needs tuning, detailed planning, and accurate practice. Such collaborations can also be impeded by regulation on the use and exchange between farms of nutrients.

11.3 Agriculture with Biodiversity

Biodiversity refers to the diversity of plants, animals, and micro-organisms, which also addresses genetic variety (within species) and variety in ecosystems. It is not solely about the presence of species but also about interactions between species within ecosystems and the coherence between ecosystems. The biodiversity in taxonomic, functional, phylogenetic facets increases from the poles to the equator. Combinations of environmental factors make this possible (Willig and Presley 2018). Herewith, the above-ground biodiversity is strongly linked to soil biodiversity, because plants and animals largely depend on the soil on or in which they live (Wall and Knox 2014). The biodiversity of plants reduces over time, and all over the world, agriculture has a large share in that loss. The biodiversity loss is also harmful to agriculture, for instance, adjustments to invasive species are needed, whereas a high biodiversity has a dampening effect on risk factors, such as climate change. Biodiversity also forms the basis for many ecosystem services (Meiresonne and Turkelboom 2014). Many economic values heavily depend on biological resources; for example, 80% of the needs of poor populations depend on them (Jones-Walters 2018). The issue is about how to align agriculture to biodiversity (Sukkel 2018).

Agrobiodiversity

Agriculture is the largest producer of biological resources for foods and it, thereby, has a vested interest in increasing biodiversity called 'Functional Agrobiodiversity (FAB)'. Biodiversity in agriculture provides several advantages for crops and animals,[3] such as making the system more robust, less sensitive to disruptions, and cost-effective. There are possibilities for realizing those advantages, particularly in organic farming with mixed cultures (Noorduyn and Sukkel 2010).

Mixed Cultures

Although the intensive growing of monocultures offers large yields, mixed cultures of two or more species have advantages (Fangou et al. 2016; Enyi 2008)—they are less susceptible to diseases and pests because species are sensitive to different growth rates, development, flowering time, and the moment of ripening. Mixed cultures also ensure a longer period of soil coverage, absorption of nutrients in soil, better micro-biotope for small animal and plant species, while more predators for harmful insects can establish themselves. Hence, mixed cultures are interesting from the point of view of sustainability.

In the practice of organic agriculture, mixed cultures of grassland and clover, of maize and field beans are therefore also used. Under a few conditions such as the right crops to combine with, the right moment to sow and harvest with good soil, mixed farming can be even more productive in terms of value and deliver better ecosystem services (Anten 2018). From the 1970s and 1980s onwards, the research shows that the yield and risks of mixed cultures depend on several factors;

[3] https://www.groenkennisnet.nl/nl/groenkennisnet/dossier/dossier-fab.htm.

such as the initial mix of the species, soil moisture, and interception of daylight by the different species depending on their leaf area and the spatial differentiation of leaves and roots because these determine of water and nutrient efficiency and consequently the growth of species (Spitters and Aerts 1983). Alternation of rows with different crops, for instance in lettuce cultivation, can also result in higher yields, less use of fertilizer and pesticides compared to the same crops in a monoculture (Fangou et al. 2016). Because different crops have different roots, leaves, and nutrient requirements, they also use soil, water, and light in different ways; and this needs labor. Possible robotization for high productivity agriculture in combination with biodiversity and good environment replaces heavy machinery and some labor on mixed cultures (Sukkel 2018). In mitigating the impacts of monocultures, mixed cultures become unavoidable; for example, the expansion of mixed cultures on small farms in sub-Saharan Africa in response to the decline in yield and soil fertility of monocultures indicates the possibilities of a win-win situation for mixed cultures from the economic and environmental perspectives (Ngwira et al. 2012).

Biodiversity in Livestock Farming

Livestock farming has intensified from the beginning of the last century. The per animal production increased due to selection and breeding programs based on the genetic predisposition for mass production and on inbreeding (Eggen 2012). Artificial insemination also enabled the spread of genetics of a single male animal over the entire cattle population. Initially, this lowered or lead to weaker kinship in local populations because siblings depended on a local supplier; but as the local markets grew into a global market and the number of genetics providers decreased—the selected genetics is increasingly used on farms. Nowadays, there are less suppliers of genetics in livestock farming but they operate globally. The drawback is that the selection increases the genetic relationship between the animals and homozygosity; which undermines the resilience of cattle (Kosowska 1992; Soerenson et al. 2005; González-Recio et al. 2007).

For example, in dairy farming in the Netherlands, the degree of kinship between bulls has risen to 12%, which has increased the chance of inbreeds. The use of genomic techniques which shortens the generation interval has contributed greatly to this (Hayes et al. 2009; de Roos et al. 2011). The effect is called 'inbreed depression'; the effects of which are a reduction in biological fitness in a population because animals are more likely to have a hereditary condition, become more susceptible to diseases and be less fertile. Health and life expectancy come under pressure, and regular corrective actions must be taken to keep the animals productive (Kardos et al. 2018). In intensive livestock farming where conditions are controlled with technical provisions, the use of medication has increased and hormones are used to keep reproduction up to a certain standard. Breeders try to reduce the possible shortcomings by influencing the genetic predisposition, but this is an additional impulse for the use of related genetics; though a good control on inbreeding helps prevent problems (Sonessen et al. 2012).

In recent years, a growing number of farms have crossbred their cattle with other breeds to reduce the degree of kinship and gain more heterozygosity in the herd.

The cross-breeding aims to combine the characteristics of different breeds in the offspring. Since the predisposition to high production that has been developed on the basis of inbreeding is often reduced, cross-breeding also reduces profitability (Smith et al. 1998). Research shows that the genetic diversity of breeds (heterosis) is positive for health, fertility, and the fitness of animals (Vanraden and Sanders 2003). The breeders also pursue varieties that fit local environmental conditions (ag4impact 2018). Attempts aiming to increase production and longevity of the dairy cattle go on because better genetic predisposition and improved operating conditions lead to economic benefits, fewer problems, and lower labor input (Devries 2013).

11.4 Soil Fertility

Soil fertility is the ability of the soil to provide a plant with water and nutrients (Schils 2012). Soils with good organic matter and an active soil life have good soil fertility, and the crops suffer less from insects. Soil fertility decreases when cultivation becomes more intensive (Su et al. 2006) often with heavy machinery and soil organic matter wil decrease and compactation of the soil will increase which on its turn reduces machinability and, increases water run-off as the storage capacity of nutrients and water decreases. That decrease causes poorer utilization of nutrients, impoverished soil life, stress in plants, and increased use of external water sources. The reduced machinability of the soil demands even heavier soil tillage machines, a larger fuel consumption and, higher CO_2 emissions. Further, reduced organic matter in the soil leads to losses of organically bound CO_2; vegetation adapts to these conditions which reduces biodiversity. Studies show relations between the losses in soil fertility, nutrient efficiency, and biodiversity (Tilman et al. 1996; Maeder et al. 2002; Dybzinski et al. 2008); the growing sensitivity of plants to diseases and pests in relation to soil fertility is because soil micro-organisms play a role in plant growth and health (Berg 2009; Miguel et al. 2003). Fertile soil also binds more carbon in the organic matter while intensive cultivation reduces the organic matter, and the removal of biomass causes a burden on the environment and climate (Balesdent et al. 2000). Improved soil fertility is made possible by optimizing supplies of the organic matter; applying crop rotation as far as possible (e.g., grass-maize-beans); working with lighter, less-intensive machines (e.g., with no-till farming); and keeping the soil covered with crops as long as possible so that the coverage can absorb nutrients, prevent outflow, and water run-off, and offers protection from some pests.

Close Cycles of Nutrients (N, P)

The closing of nutrient cycles at the farm level is rarely possible because the utilization of nutrients such as fertilizers for arable production and feedstuffs poor livestock production is inefficient. Intensive livestock production needs large external inputs of minerals, and the surplus of minerals is disposed of via animal manure. For a balance in the nutrient cycle, the quantity of livestock has to be adjusted to the arable area, or vice versa. If arable land is available, the nutrient cycle can be closed when farmers co-operate across regions because they can balance surplus versus shortage. It would require costly transport, energy, and labor, as well as crop rotation to avoid diseases, improve soil fertility, and foster biodiversity. Nevertheless, farmers can tune operations on dairy farms with ones on arable farms (de Wolf et al. 2018).

The expensive transport of manure over long distances is uneconomic and causes fraud; whereas residues of the food industries can be used for minerals and nutrients, but this requires a change in the industrial deliveries and shift from mineral fertilizers to organic ones. The circularity on distance remains unsuccessful despite many efforts, mainly because the cost of transport exceeds the benefits of circularity. Hence, most attention is paid to recycling of nutrients on farms as found in the Annual Nutrient Cycle Assessment, a tool to assess environmental performance on dairy farms (Holster 2013).

Energy and Greenhouse Gases

Modern, intensive farms use a considerable amount of energy for automated systems and machines, in barns and on land. Solar power and wind energy are profitable in many cases and farmers produce biofuels for their own use; as in the case of rapeseed oil for multi-fuel tractors. Uses of lighter machines and operations that use less energy are possible, for example, a non-turning tillage. No-till farming is a way of growing crops on soil. No-till farming also contributes to the preservation of organic matter in the soil, soil texture, CO_2 in the soil, water infiltration as well as fosters soil fertility (Nardali 2009) and reduces soil erosion caused by wind (Spears 2018). Cattle farms are sources of methane emission which is a greenhouse gas, and much research is done on methane reduction (Moe and Tyrrell 1979; Ellis et al. 2007; Gerber et al. 2013). Emissions can be reduced by better cattle feeds and increased longevity of cattle. A better longevity means less young stock is needed to replace cows, and young stock is an important source of methane. Young cows grow until they are adults, and they produce less than adult cows; therefore, efficiency per milk unit of adult cows is higher which reduces methane production per kg milk. Extending the lifespan is possible through better husbandry, management, and breeding.

11.5 Conclusions

Agriculture is the main producer of bioresources but intensive practices degrade biodiversity and soil fertility are energy consuming, polluting, wasteful, and contribute to climate change. Possibilities for more sustainable agricultural practices are accessible to many farmers. For global food production, the challenge is to develop intensive sustainable agriculture. Improvements attained in organic agriculture are considered a possible innovative alternative that can contribute to intensive sustainable agriculture. If such a shift is perceived—whether impossible or costly—technological and managerial improvements are possible. The agricultural sector has been highly innovative throughout past centuries with respectable uses of minerals, pesticides, and medicines, as well as that of animal feeds, breeds, and the fertilization of soil.

Now agriculture has to make a U-turn in the direction of fewer inputs, and a higher efficiency with respect to its environment. Much can be achieved through the dissemination of knowledge, innovations, and experimentation because the farming business adopts novelties. Sustainable agriculture is possible but requires special attention to be paid to entrepreneurship and innovations on the farm and chain levels.

References

Ag4impact (2018) https://ag4impact.org/sid/genetic-intensification/livestock-breeding/cross-breeding/

Anten NPR (2018) Functional biodiversity in crop-crop and crop-weed interactions. https://www.wur.nl/en/Research-Results/Chair-groups/Plant-Sciences/Centre-for-Crop-Systems-Analysis/MScBSc-theses/Functional-biodiversity-in-crop-crop-and-crop-weed-interactions.htm

Balesdent J et al (2000) Relationship of soil organic matter dynamics to physical protection and tillage. Soil Tillage Res 53(3–4):215–230

Berg G (2009) Plant–microbe interactions promoting plant growth and health: perspectives for controlled use of microorganisms in agriculture. Appl Microbiol Biotechnol 84(1):11–18

Campbell B et al (2017) Agriculture production as a major driver of the Earth system exceeding planetary boundaries. Ecol Soc 22(4):8

Cordell D, White S (2014) Life's bottleneck: sustaining the world's phosphorus for a food secure future. Ann Rev Env Resour 39:161–188

de Boer IJM, van Ittersum MK (2018) Circularity in agricultural production. Animal production systems and Plant production systems. Wageningen University and Research

de Goede DM (2014) Understanding robustness as an image of sustainable agriculture. PhD thesis, Wageningen University, Wageningen, NL

de Roos APW et al (2011) Effects of genomic selection on genetic improvement, inbreeding, and merit of young versus proven bulls. J Dairy Sci 94:1559–1567

de Vries W (2018) https://www.wur.nl/nl/nieuws/Het-kan-10-miljard-mensen-voeden-zonder-milieugrenzen-voor-de-aarde-te-overschrijden.htm

de Wolf K et al (2018) Verduurzaming samenwerking akkerbouw-veehouderij in Drenthe. WUR/Wageningen Plant Research, Rapport WPR-773

Devries A (2013) Cow longevity economics: the cost benefit of keeping the cow in the herd. In: Proceedings from the Cow Longevity Conference 2013, Hamra farm, Sweden

Dybzinski R et al (2008) Soil fertility increases with plant species diversity in a long-term biodiversity experiment. Oecologia 158: 85. https://doi.org/10.1007/s00442-008-1123-x

Eggen A (2012) The development and application of genomic selection as a new breeding paradigm. Animal Front 2(1):10–15

Ellis JL et al (2007) Prediction of methane production from dairy and beef cattle. J Dairy Sci 90(7):3456–3466

Enyi BAC (2008) Effects of intercropping maize or sorghum with cowpeas, pigeon peas or beans. Faculty of Agriculture, Morogoro, W., Tanzania. Experimental Agriculture 9(1):83–90

Fangou M et al (2016) Simulating potential growth in a relay-strip intercropping system: model description, calibration and testing. Centre for Crop Systems Analysis. Wageningen University, Wageningen, The Netherlands

Foley JA et al (2005) Global consequences of land use. Science 309(5734):570–574

Gerber PJ et al (2013) Tackling climate change through livestock: a global assessment of emissions and mitigation opportunities. Food and Agriculture Organisation of the United Nations (FAO)

González-Recio O et al (2007) Inbreeding depression on Female fertility and calving ease in Spanish dairy cattle. J Dairy Sci 90(12):5744–5752

Hayes BJ et al (2009) Genomic selection in dairy cattle: progress an challenges. J Dairy Sci 92(2):433–443
Holster H et al (2013) KringloopWijzer, goed geborgd!? Annual Nutrient Cycling Assessment (ANCA), adequately assured!? Rapport 676 Wageningen University and Research
Jones-Walters LM (2018) https://www.wur.nl/nl/show-longread/Biodiversiteit-longread.htm
Kalantari F et al (2017) A review of vertical farming technology: a guide for implementation of building integrated agriculture in cities. In: Advanced Engineering Forum, vol. 24 pp 76–91
Kardos M et al (2018) Genomic consequences of intensive inbreeding in an isolated wolf population. Nat Ecol Evol 2:1
Kosowska B (1992) The relationship between homozygosity level and animal physiology: iron content of plasma and whole blood as well as total iron binding capacity by transferrin (TIBC) in rats of various inbreeding coefficient. Biochem Genet 30(7/8)
Leach K (2012) Assessing the sustainability of EU organic and low input dairy farms. ORC Bulletin Nr. 111-Winter 2012
Lesschen JP (2018) https://www.wur.nl/nl/nieuws/Nutri2Cycle-naar-een-echte-kringloopeconomie.htm
Maeder P et al (2002) Soil fertility and biodiversity in organic farming. Science 296(5573):1694–1697
Meiresonne L, Turkelboom F (2014) Biodiversiteit als basis voor ecosysteemdiensten in Vlaanderen. INBO.M.2014.1817081. Instituut voor natuur—en bosonderzoek. Brussel
Miguel A et al (2003) Soil fertility management and insect pests: harmonizing soil and plant health in agroecosystems. Soil Tillage Res 72(2):203–2011
Moe PW, Tyrrell HF (1979) Methane production in dairy cows. J Dairy Sci 62(10):1583–1586
Nardali ET (2009) No-till farming: effects on soil, pros and cons and potential. https://www.researchgate.net/publication/293211592_No-till_farming_Effects_on_soil_pros_and_cons_and_potential
Ngwira AR et al (2012) On-farm evaluation of yield and economic benefit of short term maize legume intercropping systems under conservation agriculture in Malawi. Field Crops Res 132:149–157
Noorduyn L, Sukkel W (2010) Klimaatverandering te lijf. Ekoland 30 (2010)1. ISSN 0926-9142—pp 18–19
Norman D et al (1997) Defining and implementing sustainable agriculture (Kansas Sustainable Agriculture Series, Paper #1; Manhattan KS: Kansas Agricultural Experiment Station
Parr JF et al (1990) Sustainable agriculture in the United States. In: Edwards CA et al (eds) Sustainable agricultural systems. Soil and Water Conservation Society
Reytar K et al (2014) Indicators for a sustainable agriculture: a scoping analysis. World Resources Institute (WRI)
Schils R (2012) 30 vragen en antwoorden over bodemvruchtbaarheid. Wageningen Universitiy and Research, Alterra
Scholten M et al (2018) Technische briefing kringlooplandbouw. Notitie voor de vaste kamer commissie LNV. Wageningen University and Research
Smith LA et al (1998) The effects of inbreeding on the lifetime performance of dairy cattle. J Dairy Sci 81:2729–2737
Soerenson AC et al (2005) Inbreeding in Danish dairy cattle breeds. J Dairy Sci 88(5):1865–1872
Sonessen AK et al (2012) Genomic selection requires genomic control of inbreeding. Genet Sel Evol 44(1):27
Spears S (2018) https://regenerationinternational.org/2018/06/24/no-till-farming/ Advanced Engineering Forum Submitted: 2017-08-17, vol. 24, pp 76–91. ISSN: 2234-991X
Spitters CJT, Aerts R (1983) Simulation of competition for light and water in crop weed associations. Aspects Appl Biol 4:467–484
Springmann M et al (2018) Options for keeping the food system within environmental limits. Nature 562:519–525
Spruijt-Verkerke et al (2004) Duurzaamheid op biologische bedrijven. Publicatie PPR 328, Wageningen University and Research

Steinfeld H, Gerber P (2010) Livestock production and the global environment. Consume less or produce better? Proc Nat Acad Sci PNAS 107(43):18237–18238. https://doi.org/10.1073/pnas.1012541107

Su YZ et al (2006) Long-term effect of fertilizer and manure application on soil-carbon sequestration and soil fertility under the wheat–wheat–maize cropping system in northwest China. Nutr Cycl Agroecosyst 75:285. https://doi.org/10.1007/s10705-006-9034-x

Sukkel W (2018) https://www.wur.nl/nl/show-longread/Biodiversiteit-longread.htm

Tilman D et al (1996) Productivity and sustainability influenced by biodiversity in grassland ecosystems. Nature 379:718–720

UDV (2018) http://www.uitvoeringsagendaduurzameveehouderij.nl/

Vanraden PM, Sanders AH (2003) Economic merit of crossbred and purebred US dairy cattle. J Dairy Sci 86(3):1036–1044

Wall DH, Knox MA (2014) Soil Biodiversity. Reference Module in Earth Systems and Environmental Sciences, pp 136–141

Willig MR, Presley SJ (2018) Latitudinal gradients of biodiversity: theory and empirical patterns. In: Encyclopedia of the anthropocene

Chapter 12
Potential Applications of Lignin

Marlene Kienberger

The increasing consumption of fossil fuels is associated with environmental issues such as global warming and environmental pollution. Therefore, efforts have been made to find sustainable alternatives for different materials and energy sources. Plant biomass, especially wood, is the most important renewable material. The main constituents of vascular plants are cellulose with a content of 42–51%, hemicellulose with a content of 24–40% and lignin with a content of 18–30%. Cellulose and hemicellulose represent the cell walls, while lignin acts as a strengthener thereof, and inhibits enzymatic degradation, regulates water transport, and demonstrates antibacterial activity. Cellulose is used for paper. The focus, herewith, is on lignin. Lignin is mainly used for fuel but has many more valuable applications which are introduced in this chapter.

12.1 Source of Lignin

During the digestion process in wood pulping, lignin and hemicellulose are dissolved into cooking liquor; cellulose remains solid and is further processed to different paper products. Cellulose is the main product in the pulp and paper industry, and lignin and hemicellulose though by-products are crucial in the chemical recovery cycle for energy utilization. The heating value of lignin amounts to 23.4 MJ/kg, and hence, lignin is an appropriate fuel for the chemical recovery cycle. Due to the use of lignin as a biofuel in the chemical recovery cycle, fossil fuels play a minor role or are not even needed for the production of paper.

There are two main digesting processes in the pulp and paper industry. The leading process is the Kraft process wherein sodium hydroxide and sodium sulfide are used to gain cellulose fibers. The cooking liquid (black liquor) of this process contains Kraft

M. Kienberger (✉)
Graz University of Technology, Graz, Austria
e-mail: marlene.kienberger@tugraz.at

Y. Krozer and M. Narodoslawsky (eds.), *Economics of Bioresources*,
https://doi.org/10.1007/978-3-030-14618-4_12

lignin, a complex biopolymer with various functional groups, e.g., methoxyl groups, phenolic hydroxyl groups, and aromatic rings. The second processing method is the sulfite process. Here, the wood chips react with magnesium sulfite which leads to the formation of lignosulfonates in the cooking liquid. Lignosulfonates exhibit a similar structure as lignin, but with sulfonate groups in their aliphatic side chains. Altogether around 55 million tons of lignin are produced during wood cooking worldwide, of which, about 98% are used for energy (Behr and Seidensticker 2018). Only a small amount of lignin is used in high value products and materials; though the interest for such uses is on the rise because lignin is produced in large quantities and has numerous interesting properties, such as—biodegradability, high thermal stability, high carbon content, and antioxidant activity (Kai et al. 2016; Saake and Lehnen 2012).

12.2 Applications of Lignosulfonates

Existing Applications

Lignosulfonates in Expanders for Lead/Acid Batteries

Lead-acid batteries lose performance because of structural changes and passivation of the active material. Expanders are additives which improve performance by reducing the negative effect of material in batteries, effectively prolonging its cycle life, and capacity. Modern expanders typically consist of barium sulfate, carbon black, and lignin derivatives, for instance, lignosulfonates (Boden 1998). Lignosulfonates in such expanders improve charge acceptance, high temperature, and cold-cranking performance; prolongs durability and increases the load in the cycle life (Boden 1998; Rand et al. 2002; With and Boden 2008; Blecua et al. 2017). Lignosulfonates containing expanders are used in automotive batteries as well as industrial batteries (Boden 1998).

Lignosulfonates as a Flocculant and an Emulsion Stabilizer

Lignosulfonates contain hydrophobic and hydrophilic properties, therefore acting as flocculants and emulsion stabilizers, depending on its molecular weight. Chan et al. (1976) reported that the high molecular weight lignosulfonates act as dispersants, and the low molecular weight lignosulfonates as flocculants; which suggests that the flocculant properties of lignosulfonates increase with increase in molecular weight (Sjoblom 2001; Chan et al. 1976). However, various parameters influence the flocculant behavior of lignosulfonates, such as their concentration, pH-value, or the presence of specific chemicals (Sjoblom 2001; Chan et al. 1976; Gundersen and Sjoblom 1991). Therefore, Chan et al. reported that a more effective flocculant is obtained by mixing lignosulfonates with other materials, for example, alum (Chan et al. 1976).

Lignosulfonates as a Water Reducing Agent and Plasticizer in Concrete

The application of lignosulfonates as a water reducing agent in concrete started in the 1930s, when unmodified waste liquor from the sulfite process was added to the cement mixture. A dry waste liquor content of 0.15% per wet ton reduced the water content by 10–12% (Scripture and Heights 1937; Reknes 2013). Nowadays, the uses of lignosulfonates are refined and modified before their applications (Reknes 2013). The plasticizing effect of lignosulfonates enhances flow-ability and pour-ability, which are other reasons for using lignosulfonates in concrete. Even though synthetic compounds show better performances when compared to lignosulfonates, the latter has the advantage of being renewable and cheaper (Plank 2004). Furthermore, by using lignosulfonates in concrete, a considerable amount of CO_2 is saved (Reknes 2013).

Lignosulfonates as a Dust Control Agent

Unpaved surfaces of infrastructure are the main reasons for dust on roads and from construction sites. Since the 1920s, lignosulfonates have been widely used as a dust controlling agent. Compared to chloride salts and petroleum-based materials, lignosulfonates are cheaper, renewable, and non-toxic. Spraying lignosulfonate solution, diluted with water to about 25% solid content (Applications 1992), onto roads leads to a higher stability of the surface over time. Furthermore, it traps dust particles, plugs pores, and reduces water penetration (Applications 1992; Calvo-Flores et al. 2015).

Potential Applications of Lignosulfonates

Metal Adsorbents

The functional groups of lignosulfonates make them attractive as an adsorption agent, for example, for metal ions (Yang et al. 2014a). It is suggested that a metal complex is formed, where the carboxylic and sulfonate groups in the lignin structure act as complexation aids (Aro and Fatehi 2017). Yang et al. synthesized a lignosulfonate–graphene oxide–polyaniline nanocomposite with an adsorption capacity of 216.4 mg per gram for Pb(II) ions at 30 °C, which offers great potential for the removal of Pb(II) from wastewater (Yang et al. 2014a). In another study, Luo and Lü synthesized a lignosulfonate–polypyrrole composite by in situ polymerization which shows good adsorption potential for the removal of Ag(I) and Pb(II) from water (Luo and Lü 2015).

Production of Vanillin

The flavor vanillin was produced from lignosulfonates in the past, but is substituted by the production of vanillin from guaiacol and glyoxylic acid (Bjørsvik and Minisci 1999; Salvesen et al. 1948; Tarabanko et al. 2004). Since lignosulfonates are an abundant biogenic by-product in the pulp and paper industry, the synthesis of vanillin from lignosulfonates has gained more interest in recent years (Bjørsvik and Minisci 1999; Tarabanko et al. 2004). The main hindrance for its industrial application is a low yield, of 4–6% vanillin synthesized from lignosulfonates (Bjørsvik and Minisci 1999).

Crop Protection and an Additive in Fertilizers

Due to its biodegradability, lignosulfonate has gained interest as an additive in crop protection agents or fertilizers. There are different reasons for incorporating lignosulfonates in such products, for instance to obtain a controlled release of active substances (Detroit 1988), or harden and improve the anti-cracking properties of fertilizer particles (Detroit 1991). In addition, it is observed that lignosulfonates enhance the efficiency of fertilizers although the reason for this effect is not yet known (Meier et al. 1993).

Additive in Packaging Materials

Several functional groups in the lignosulfonate structure offer a remarkable ability to enhance the properties of materials. Johansson et al. investigated a coating based on the aqueous dispersion of clay, latex, starch, lignosulfonates, and laccase. The oxidation of lignosulfonates through laccase forms radicals which then polymerize (radical coupling). Measurements performed on this material show that active enzymes remain, even after drying at temperatures higher than 100 °C. The function as oxygen scavenger requires a high relative humidity, which makes this coating suitable for packaging of high-moisture foods (Johansson et al. 2012).

12.3 Applications of Kraft Lignin

Existing Applications

Kraft Lignin as Fuel

The main equipment in an integrated biorefinery of pulp and paper production is the recovery boiler, where the process chemicals Na_2S and NaOH are recovered by burning the heavy black liquor. The combustion of black liquor produces steam, which is used for the production of electricity at the paper machine on-site. The main contribution to the heating value of black liquor comes from lignin. A Kraft pulping mill with a production of 1500 t/d of black liquor can produce 25–35 MW of electricity (Tran and Vakkilainnen 2012). Increasing demand for various paper products leads to an increase in production. As the pulp production increases, the heating value of the heavy black liquor increases. The recovery boiler used in the chemical recovery cycle reaches, in the meanwhile, its inherent design limit due to the high calorific load that is delivered by the high lignin load. Therefore, lignin can be isolated from the black liquor. State-of-the-art pulp and paper processes involve lignin precipitation. The isolated lignin may be sold or used, e.g., as fuel for the lime kiln on-site (Kouisni et al. 2011). Hence, black liquor ranks among one of the most important fuels in the world and is the largest biofuel (Tran and Vakkilainnen 2012).

Potential Applications

Lignin in Phenol-Formaldehyde Resins

Phenol-formaldehyde (PF) resins are synthetic polymers synthesized from fossil fuel and find use in various technical applications due to their magnificent mechanical strength, resistance against moisture, and heat. Lignin is considered a phenol substituent in PF resins (Kouisni et al. 2011; Kai et al. 2016; Chen et al. 2013), due to environmental issues, a turn into a bioeconomy and the reduction in production costs. The main issues in the application of lignin are its variable composition and low reactivity. Attempts to improve the latter have been made by conducting different chemical modification processes such as phenolation, demethylation, and methylolation (Kai et al. 2016; Yang et al. 2014b; Olivares et al. 1988; Gonçalves and Benar 2001; Vázquez et al. 1997). Various studies have been performed on incorporating lignin into PF resins for plywood production. Experiments showed that up to 30% phenol can be replaced by lignin without causing major changes in the properties of plywood (Kouisni et al. 2011; Kai et al. 2016). Other studies show that a lignin substitution of up to 60% is possible, which simultaneously increases the adhesive strength of plywood (Kai et al. 2016; Qiao et al. 2015).

Lignin-Based Carbon Fibers

Carbon fiber materials are regularly used for energy in aircrafts, aerospace, and wind mills. However, high costs of production limit their wider use. The production of the precursor is especially expensive (Mainka et al. 2015; Kadla et al. 2002). A potential alternative to currently used precursors is lignin. Kadla et al. (2002) reported a method to produce carbon fiber from Kraft lignin using a thermal extrusion process (Kadla et al. 2002). Another reported method is melt-spinning of lignin into fibers under inert conditions followed by oxidative thermo stabilization (Baker et al. 2012). This is a promising way to decrease the emission of cars by reducing their weight using carbon fiber reinforced plastic (CFRP) made of lignin (Mainka et al. 2015).

Lignin as Asphalt Binder

Asphalt (or bitumen) is a petroleum-derived material. Its demand grows constantly due to the expansion and maintenance of roads (Xie et al. 2017; Bourzac 2015). The substitution of asphalt with natural compounds is researched worldwide. Due to its similar structure, lignin emerges as an interesting option. In order to replace asphalt, new bioresources should offer a similar or better performance during high temperatures, e.g., prevention of melting and deformation; and during low temperatures, e.g., low cracking potential, just as conventional asphalt would. Xie et al. report that raw Kraft lignin has to be processed in order to meet the requirements. Therefore, Kraft lignin is fractionated into a water-soluble fraction and a water-insoluble one, using a laccase-mediator system. The water-soluble fraction exhibited better performance during high and low temperatures, in comparison with asphalt. Xie et al. suggested that lignin possibly acts as a cross-linker with the asphaltenes through dipolar–dipolar interactions. Furthermore, hydroxyl groups in the lignin structure would form hydrogen bonds. In another series of experiments, Xie et al. used formic acid, iron, and H_2O_2 to modify Kraft lignin. The result was that the insoluble fraction of the chemically modified lignin could work as good high-quality asphalt binder, although the performance was not as good as that of the soluble fraction made by biologically processed lignin. All experiments were carried out with a lignin content of up to 20% (Xie et al. 2017). Van Vliet et al. (2016) modified the lignin by blocking the hydroxyl groups to make it more hydrophobic. The blending experiments showed that fractions of up to 25% can be mixed with used bitumen for recycling (Van Vliet et al. 2016).

Production of Vanillin

Vanillin (the major flavoring in vanilla) is used as a flavor agent, perfume additive, or chemical precursor in the pharmaceutical industry. In addition to the extraction of vanillin from its natural bioresource, vanilla, a large amount is produced by chemical synthesis (Shakeri et al. 2009). Lignin has gained attention in producing vanillin from natural sources. The production of vanillin by oxidation of sulfite-lignin is already being commercially applied (Shakeri et al. 2009; Bryan 1954; Fargues et al. 1996). The use of Kraft lignin for vanillin production, using oxidation under alkaline conditions, to achieve yields that vary from 2.8 to 10.8% of wet ton vanillin per wet ton lignin is already underway (Shakeri et al. 2009; Fargues et al. 1996; Rodrigues 1995; Fache et al. 2016; da Silva et al. 2009).

Antimicrobial Agent

Due to its phenolic groups or—more precisely its side chains and functional groups—lignin exhibits antimicrobial behavior (Kai et al. 2016; Nada et al. 1989). Nada et al. (1989) demonstrated that the antimicrobial effect of lignin highly depends on the extraction temperature. Too high a temperature leads to a loss of antimicrobial activity (Nada et al. 1989). There are various possible applications of lignin as an antimicrobial agent, such as its use in tissue engineering for biomedical implants and other antibacterial materials (Kai et al. 2016). Furthermore, it was shown that lignin is able to improve the efficiency of silver nanoparticles (Kai et al. 2016; Richter et al. 2015).

Smart Materials

Materials that can sense their environment, and accordingly react to the stimuli of moisture, temperature, load, or pH are called smart materials (Kai et al. 2016). Feng et al. (2011) developed a temperature-sensitive, lignin-based hydrogel through graft polymerization of acetic acid and lignin with poly (N-isopropylacrylamide) (Kai et al. 2016; Feng et al. 2011). Dallmeyer et al. (2013) prepared the lignin-based moisture-responsive fabrics by the co-electrospinning of Kraft lignin (Kai et al. 2016; Dallmeyer et al. 2013). Furthermore, Li et al. (2015) explored the fabrication of a thermally simulated shape memory co-polymeric elastomer from lignin (Li et al. 2015). However, it is not always clear as to what extent lignin is responsible for the respective behavior of various materials.

12.4 Uses of Modified Lignin

Mechanical Reinforcing Composites

Natural lignin acts as a mechanical support in plants. Thus, scientists have investigated the application of lignin as a reinforcing agent in various materials (Kai et al. 2016). Kai et al. (2016) summarized various attempts of blending polymers with modified lignin; for example, alkylated lignin and esterified lignin. Some experimental results show that the addition of lignin can improve the properties of the final products, for instance, rubber–lignin composites and thermoplastics. However, the addition of lignin is often accompanied by drawbacks such as a significant drop in elongation (Kai et al. 2016).

Packaging Material

Lignin exhibits antioxidant activity due its phenolic structure (Kai et al. 2016; Domenek et al. 2013). Kai et al. (2016) summarized several studies which have shown that lignin can significantly increase the oxygen induction time, even when applied in small amounts (~1% wet ton) (Kai et al. 2016). Other studies involve the coating of packaging material and paperboard with esterified lignin. Again, a significantly lower oxygen transmission rate was observed (Hult et al. 2013; Vartiainen et al. 2014). This might be a promising way to produce more efficient and sustainable packaging material.

UV-Active Materials

During polymerization of monolignols, UV chromophoric groups are generated. Thus, lignin shows absorption ability in the UV and visible region (Kai et al. 2016; Chaochanchaikul et al. 2012). Hambardzumyan et al. (2012) successfully synthesized nanocomposite coatings consisting of cellulose nanocrystals and lignin. The nanocrystals were prepared without functionalization or chemical modification of the two biopolymers. The composite system obtained performed well with regards to UV blocking. Furthermore, the material was highly transmitting within the visible spectrum (Kai et al. 2016; Hambardzumyan et al. 2012). Qian et al. investigated the incorporation of lignin into various creams and sunscreens. It was found that lignin led to an increase in Sun Protection Factor (SPF), in proportion to the concentration of lignin. The addition of only 2% wet ton lignin to a sunscreen with a SPF of 15 enhanced its sunscreen effect to that of one with SPF of 30 (Kai et al. 2016; Qian et al. 2015).

Utilization in Thermoplastic and Polymer Blends

Due to its chemical structure, lignin is considered a substituent for non-renewable compounds in polymer blends and thermoplastics. To overcome the problem of lignin's poor miscibility in the blends and solvents, certain methods have been tried, they are the—addition of coupling agents (e.g., ethylene-vinyl acetate) or chemical modification of lignin (e.g., esterification) (Dehne et al. 2016; Pawar et al. 2016;

Luo et al. 2017; Koivu et al. 2016; Thielemans and Wool 2005). Another point to be addressed is the diversity of the molecular weight of lignin. Saito et al. (2012) reported that high molecular weight lignin obtained by washing with methanol, or reaction with formaldehyde led to the formation of a more continuous networked structure in the synthesized thermoplastics (Saito et al. 2012). Results have shown that the incorporation of lignin, especially modified lignin, into thermoplastic and polymer blends offers promising ways to enhance the biodegradability of various materials.

12.5 Conclusions

The various possibilities for high value uses of lignin are presented. Lignin, a major component of wooden bioresources, is mainly used as biofuel in the pulp and paper industries. Since the resources for biofuels are widely available in these industries, alternative uses are investigated. Presently, the main uses are expanders in lead-acid batteries, flocculants, emulsion stabilizers, and water reducing agents. The potential uses are the replacement of phenol-formaldehyde (PF) resins, carbon fibers in aircrafts, chemical vanillin, as well as uses for antimicrobe agents and thermal sensors. Modification of lignin is researched for the reinforcement of materials, coating of paper and board in packaging, UV protectors, and degradable thermoplastic. Many possibilities for the use of lignin are investigated; however, what remains to be defined are options that would be cost-effective when compared to rival resources, mainly based on the synthesizing of mineral-based compounds.

References

Applications R (1992) Lignins: a safe solution for roads 1(3)

Aro T, Fatehi P (2017) Production and application of lignosulfonates and sulfonated lignin. ChemSusChem, 1861–1877

Baker DA, Gallego NC, Baker FS (2012) On the characterization and spinning of an organic-purified lignin toward the manufacture of low-cost carbon fiber. J Appl Polym Sci 124(1):227–234

Behr A, Seidensticker T (2018) Einführung in die Chemie nachwachsender Rohstoffe. Springer Spektrum, Berlin, Heidelberg

Bjørsvik H-R, Minisci F (1999) Fine chemicals from lignosulfonates. 1. Synthesis of vanillin by oxidation of lignosulfonates. Org Process Res Dev 3(5):330–340

Blecua M, Fatas E, Ocon P, Valenciano J, de la Fuente F, Trinidad F (2017) Influences of carbon materials and lignosulfonates in the negative active material of lead-acid batteries for microhybrid vehicles. J Energy Storage 11:55–63

Boden DP (1998) Selection of pre-blended expanders for optimum lead/acid battery performance. J Power Sources 73(1):89–92

Bourzac K (2015) Inner workings: paving with plants. Proc Natl Acad Sci USA 112(38):11743–11744

Bryan CC (1954) Manufacture of vanillin from lignin. United States Patent, 2,692,291

Calvo-Flores FG, Dobado JA, Isac-GarcÃa JÃ, MartÃn-MartÃnez FJ (2015) Lignin and lignans as renewable raw materials: chemistry, technology and applications. Wiley
Chan SLH, Baker CGJ, Beeckmans JM (1976) Flocculating properties of high molecular weight lignosulphonates. Powder Technol 13:223–230
Chaochanchaikul K, Jayaraman K, Rosarpitak V, Sombatsompop N (2012) Influence of lignin content on photodegradation in wood/HDPE composites under UV weathering. BioResources 7(1):38–55
Chen Y, Frihart CR, Cai Z, Lorenz LF, Stark NM (2013) Lignin-based phenol-formaldehyde resins from purified CO_2 precipitated Kraft lignin (PCO2KL). In: International conference on wood adhesives, pp 601–611
da Silva EAB, Zabkova M, Araújo JD, Cateto CA, Barreiro MF, Belgacem MN, Rodrigues AE (2009) An integrated process to produce vanillin and lignin-based polyurethanes from Kraft lignin. Chem Eng Res Des 87(9):1276–1292
Dallmeyer I, Chowdhury S, Kadla JF (2013) Preparation and characterization of Kraft lignin-based moisture-responsive films with reversible shape-change capability. Biomacromol 14(7):2354–2363
Dehne L, Vila Babarro C, Saake B, Schwarz KU (2016) Influence of lignin source and esterification on properties of lignin-polyethylene blends. Ind Crops Prod 86:320–328
Detroit WJ (1988) Controlled release formulation for fertilizers. United States Patent, 4,756,738
Detroit WJ (1991) Lignosulfonates treated fertilizer particles. United States Patent, 5,041,153
Domenek S, Louaifi A, Guinault A, Baumberger S (2013) Potential of lignins as antioxidant additive in active biodegradable packaging materials. J Polym Environ 21(3):692–701
Fache M, Boutevin B, Caillol S (2016) Vanillin production from lignin and its use as a renewable chemical. ACS Sustain Chem Eng 4(1):35–46
Fargues C, Mathias Á, Rodrigues A (1996) Kinetics of vanillin production from Kraft lignin oxidation. Ind Eng Chem Res 35(1):28–36
Feng Q, Chen F, Wu H (2011) Preparation and characterization of a temperature-sensitive lignin-based hydrogel. BioResources 6(4):4942–4952
Gonçalves AR, Benar P (2001) Hydroxymethylation and oxidation of organosolv lignins and utilization of the products. Biores Technol 79(2):103–111
Gundersen SA, Sjoblom J (1991) High and Low-molecular-weight lignosulfonates and Kraft lignins as Oil/water-emulsion stabilizers studied by means of electrical conductivity. Colloid Polym Sci 277:462–468
Hambardzumyan A, Foulon L, Chabbert B, Aguié-Béghin V (2012) Natural organic UV-absorbent coatings based on cellulose and lignin: designed effects on spectroscopic properties. Biomacromol 13(12):4081–4088
Hult EL, Koivu K, Asikkala J, Ropponen J, Wrigstedt P, Sipilä J, Poppius-Levlin K (2013) Esterified lignin coating as water vapor and oxygen barrier for fiber-based packaging. Holzforschung 67(8):899–905
Johansson K, Winestrand S, Johansson C, Järnström L, Jönsson LJ (2012) Oxygen-scavenging Coatings and Films Based on Lignosulfonates and Laccase. J Biotechnol 161(1):14–18
Kadla JF, Kubo S, Venditti RA, Gilbert RD, Compere AL (2002) Lignin-based carbon fibers for composite fiber applications. Carbon 40:2913–2920
Kai D, Tan MJ, Chee PL, Chua YK, Yap YL, Loh XJ (2016) Towards lignin-based functional materials in a sustainable world. Green Chem 18(5):1175–1200
Koivu KAY, Sadeghifar H, Nousiainen PA, Argyropoulos DS, Sipilä J (2016) Effect of fatty acid esterification on the thermal properties of softwood Kraft lignin. ACS Sustain Chem Eng 4(10):5238–5247
Kouisni L, Fang Y, Paleologou M, Ahvazi B, Hawari J, Zhang Y, Wang X-M (2011) Kraft lignin recovery and its use in the preparation of lignin-based phenol formaldehyde resins for plywood. Cellul Chem Technol 45(7–8):515–520

Li H, Sivasankarapillai G, McDonald AG (2015) Highly biobased thermally-stimulated shape memory copolymeric elastomers derived from lignin and glycerol-ADIPIC acid based hyperbranched prepolymer. Ind Crops Prod 67:143–154
Luo J-J, Lü Q-F (2015) Controllable preparation and heavy-metal-ion adsorption of lignosulfonate-polypyrrole composite nanosorbent. Polym Compos 36(8):1546–1556
Luo S, Cao J, McDonald AG (2017) Esterification of industrial lignin and its effect on the resulting poly(3-hydroxybutyrate-co-3-hydroxyvalerate) or polypropylene blends. Ind Crops Prod 97:281–291
Mainka H, Täger O, Körner E, Hilfert L, Busse S, Edelmann FT, Herrmann AS (2015) Lignin—an alternative precursor for sustainable and cost-effective automotive carbon fiber. J Mater Res Technol 4(3):283–296
Meier JN, Fyles JW, MacKenzie AF, O'Halloran IP (1993) Effects of lignosulfonate-fertilizer applications on soil respiration and nitrogen dynamics. Can J Soil Sci 73(2):233–242
Nada AMA, El-Diwany AI, Elshafei AM (1989) Infrared and antimicrobial studies on different lignins. Acta Biotechnol 9(3):295–298
Olivares M, Guzmán JA, Natho A, Saavedra A (1988) Kraft lignin utilization in adhesives. Wood Sci Technol 22(2):157–165
Pawar SN, Venditti RA, Jameel H, Chang H-M, Ayoub A (2016) Engineering physical and chemical properties of softwood Kraft lignin by fatty acid substitution. Ind Crops Prod 89:128–134
Plank J (2004) Applications of biopolymers and other biotechnological products in building materials. Appl Microbiol Biotechnol 66(1):1–9
Qian Y, Qiu X, Zhu S (2015) Lignin: a nature-inspired sun blocker for broad-spectrum sunscreens. Green Chem 17(1):320–324
Qiao W, Li S, Guo G, Han S, Ren S, Ma Y (2015) Synthesis and characterization of phenol-formaldehyde resin using enzymatic Hydrolysis lignin. J Ind Eng Chem 21:1417–1422
Rand DAJ, Boden DP, Lakshmi CS, Nelson RF, Prengaman RD (2002) Manufacturing and operational issues with lead-acid batteries. J Power Sources 107(2):280–300
Reknes IK (2013) Quality lignosulfonate for concrete. The Masterbuilder, 84–90
Richter AP, Brown JS, Bharti B, Wang A, Gangwal S, Houck K, Cohen Hubal EA, Paunov VN, Stoyanov SD, Velev OD (2015) An environmentally benign antimicrobial nanoparticle based on a silver-infused lignin core. Nat Nanotechnol 10(9):817–823
Rodrigues AE (1995) Production of vanillin by oxidation of pine Kraft lignins with oxygen. Holzforschung 49(3):273–278
Saake B, Lehnen R (2012) Lignin, Ullmann's encyclopaedia of industrial. Chemistry 21:21–36
Saito T, Brown RH, Hunt MA, Pickel DL, Pickel JM, Messman JM, Baker FS, Keller M, Naskar AK (2012) Turning renewable resources into value-added polymer: development of lignin-based thermoplastic. Green Chem 14(12):3295–3303
Salvesen JR, Brink DL, Diddams DG, Owzarski P, Owzarski W (1948) Process for making vanillin. United States Patent Office, 2,434,626
Scripture EW, Heights S (1937) Indurating composition for concrete. United States Patent Office, 2,081,642
Shakeri A, Rad SM, Ghasemian A (2009) Oxidative production of vanillin from industrial lignin using oxygen and nitrobenzene: a comparative study. Int J Allied Sci 2(24):1165–1171
Sjoblom J (2001) Encyclopedic handbook of emulsion technology. CRC Press
Tarabanko VE, Petukhov DV, Selyutin GE (2004) New mechanism for the catalytic oxidation of lignin to vanillin. Kinet Catal 45(4):569–577
Thielemans W, Wool RP (2005) Lignin esters for use in unsaturated thermosets: lignin modification and solubility modeling. Biomacromol 6(4):1895–1905
Tran H, Vakkilainnen EK (2012) The Kraft chemical recovery process. TAPPI Kraft Recovery Course, pp 1–8
Van Vliet D, Slaghek T, Giezen C, Haaksman I (2016) Lignin as a green alternative for bitumen. Proceedings of E&E congress 2016—6th Eurasphalt Eurobitume Congress

Vartiainen J, Vähä-Nissi M, Harlin A (2014) Biopolymer films and coatings in packaging applications—a review of recent developments. Mater Sci Appl 5(10):708–718
Vázquez G, González J, Freire S, Antorrena G (1997) Effect of chemical modification of lignin on the gluebond performance of lignin-phenolic resins. Biores Technol 60(3):191–198
With LBE, Boden DP (2008) Lead-acid battery expanders with improved life at high temperatures. United States Patent Application Publications, US2008/0305396 A1, 1, (19)
Xie S, Li Q, Karki P, Zhou F, Yuan JS (2017) Lignin as renewable and superior asphalt binder modifier. ACS Sustain Chem Eng 5(4):2817–2823
Yang J, Wu JX, Lü QF, Lin TT (2014a) Facile preparation of lignosulfonate-graphene oxide-polyaniline ternary nanocomposite as an effective adsorbent for Pb(II) ions. ACS Sustain Chem Eng 2(5):1203–1211
Yang S, Wen J-L, Yuan T-Q, Sun R-C (2014b) Characterization and phenolation of biorefinery Technical lignins for lignin–phenol–formaldehyde resin adhesive synthesis. RSC Adv 4(101):57996–58004

Chapter 13
Bioresources for Cultural Services

Yoram Krozer and Maia Lordkipanidze

Bioresources contribute to the regulation of environmental qualities, provide energy, food, other materials; deliver cultural values that encompass basic resources for scientific, artistic, and educational inspiration; support esthetical, ethical, and spiritual behavior, as well as foster tranquility, relaxation, leisure, quality of life, and suchlike amenities. Although the consumption of bioresources is usually associated with the uses of materials, the consumption value of the non-material activities in cultural ecosystem services is fast growing in terms of scientific interest and economic development. They refer to spiritual enrichment, cognitive development, reflection, leisure and recreation, as well as ethical and esthetic experiences. Herewith, the uses of bioresources for health, tourism, and education are introduced with a focus on urban areas where most people in the world live.

13.1 Bioresources for Risk Prevention

Urban areas accommodate a large volume of bioresources in parks, gardens, orchards, playgrounds, and such vegetation is often rich in biodiversity. Their primary purpose is for health and well-being, though they also provide materials such as biofuels. Bioresources foster healthy living via the natural environment. Trees deliver shade and evaporation which have a cooling effect. That cooling effect is particularly important in dense urban areas where 'heat islands' cause health risks. Plants also contribute to flood prevention as they temper water flow, create spongy soil, and break waves along with leisure and other cultural

Y. Krozer (✉)
University of Twente - CSTM, Enschede and Sustainable Innovations Academy, Amsterdam, The Netherlands
e-mail: krozer@xs4all.nl

M. Lordkipanidze
University of Twente, Enschede, The Netherlands

Y. Krozer and M. Narodoslawsky (eds.), *Economics of Bioresources*,
https://doi.org/10.1007/978-3-030-14618-4_13

ecosystem services; combinations of the regulatory and cultural ecosystem services emerges, labeled as 'building with nature'. The uses of constructed wetlands, vegetation roofs, wastewater gardens, and such vegetation for reduction in air and water pollution in combination with scientific, education, and esthetic services also disseminate. Further, in health care, tourism and education there are more direct uses of bioresources as the cultural ecosystem services.

13.2 Health Care

Preventive Health Care in Urban Areas

Inquiries show that people associate bioresources with well-being and health; well-being indicated as 'satisfied with life' which is a key factor of perception for healthy living (Veeneklaas et al. 2011). Citizens demand more bioresources in urban areas in relation to outdoor activities which are considered important for healthy living; for instance, factors that reduce obesity and stress (Godbey et al. 2005). Indeed, living nearby bioresources in dense urban areas encourages outdoor activities. More frequent outdoor activities in turn correlate with the prevention of health risks within the urban areas (Triguero-Mas et al. 2017). For example, correlations between obesity among children and living in the vicinity of parks suggest that obesity decreases by about 15% among children who live close to these bioresources. This link is attributed to a greater frequency of outdoor activities when people live nearby bioresources. The vicinity of bioresources in neighborhoods is important for the outdoor activities, it means high density of parks, gardens, and such vegetation in urban areas; it is not per se the scale of vegetation. Social interactions drive the outdoor activities (Vreke et al. 2006). Large parks also encourage outdoor activities, but are less effective because they generate fewer interactions. Many distributed areas of vegetation are recommended for preventive health care.

Assessment of the uses of bioresources for outdoor activities and prevention of health risks in a few dozen cities of the Netherlands also reveals high correlations. Estimates show that greater access to bioresources in the urban areas facilitate outdoor activities which reduce the risk of cardiovascular disease, depression, neck and back pain, asthma, diabetes, phobias, lung diseases, urine infections, migraine, and intestine infections. Based on these correlations, social costs and benefits of additional bioresources in Dutch cities are estimated (KPMG 2012). Table 13.1 summarizes the costs and benefits.

Given that €6.3 billion was the expenditure on these diseases in 2014—it means €374 per person was spent on bioresources-related diseases that year. It is estimated that an additional 10% of bioresources can generate a net social benefit, even if only a few percent of the healthcare costs related to those diseases are reduced; this decrease is in addition to lower job absenteeism and less visits to healthcare facilities. The annual social benefit is about €11 per person and annual cost is about €8 per person, including an additional 20% cost for small-scale parks distributed across cities. The assessment is plausible though it cannot show causalities between preventive health care and bioresources, because sound theory about the relation between bioresources and health is incomplete and empirical data refers to one country only. Even though

Table 13.1 Cost-benefit analyses of 10% additional bioresource in cities

Benefit	Units	Amount
People in the Netherlands	Million	17
Curative expenditures	€ million	6300
Assumed curative decrease	Percent	3%
Annual social benefits	€/person/year	11
Cost		
Norm bioresource per household	m^2/year	75
10% for 2.1 person/household	m^2/person	4
Investment	€/m^2	41
Depreciation	€/m^2/year	4
Maintenance	€/m^2/year	3
Annual costs, 20% extra for high distribution	€/m^2/year	8
Payback	Months	9

the correlations between outdoor activities and health risks do not disclose relations, they illustrate bioresources as cultural services (D'Alessandro et al. 2015).

Curative Health Care

Parks, gardens, and pastoral life are often considered contributors to recovery from ills (Maller et al. 2002). The uses of bioresources for curative purposes refer to a sensing of nature. The seminal paper of Ulrich (1984) describes an experiment set in a hospital with 23 patients recovering from surgery; it elucidated that patients with a view of the garden recovered faster than those without a garden view (Ulrich 1984). Several subsequent studies on the passive uses of bioresources confirm such positive effects on the duration of hospitalization. The explanation is that, the view of bioresources contributes to the psychological restoration of patients, as it reduces post-surgery stress. Factors that foster such a recovery are also explored, but the mechanisms are unclear (Nieuwenhuijsen et al. 2014). The hypothesis is that the perception of nature during daylight contributes to restoring the biological rhythm of patients, which enhances their physical recovery; the disturbance of the biological rhythm is a well-known and established cause of malfunctioning.

The use of bioresources during the process of cure in hospitals can be increased by changes in the design and construction of hospitals and organization of health care. Such changes are often net beneficial because the introduction of bioresources is rarely costly whereas the duration of cure is a major cost factor when admitted in hospital. Design and construction of hospitals with more access to bioresources would reduce the costs of hospitals, and thereby the insurance costs as well (Beute and de Kort 2013). The process of cure in hospitals could also be shortened by a longer recovery period outside hospitals at a location with access to bioresources. The recovery can be enhanced with information technologies that stimulate outdoor activities; for instance, serious gaming with videos and sensors which motivate patients to be active in nature. Such changes contribute to cost-savings in health care.

13.3 Tourism

A large, growing market of biodiversity cultural services is tourism. Since leisure time has increased with the shortening of working hours and increase in life expectancy, activities in destinations outside home and work are on the rise, referred to as 'tourism'. In the period between 1995 and 2015, the working hours of an individual has shortened by about 0.4% the annual average in high income countries (OECD). Concurrently, the average life span has grown by 0.3%, which cumulatively amounts to 23% more leisure time for tourism, among others. The average annual real income per capita has also grown during that period of time, by roughly 1%, which affected a cumulative increase of 35% in real income (Krozer 2015). These figures are presumably higher on the global scale indicated by the fact that leisure time and income per capita grew faster in lower income countries. In effect, many more people pursue domestic and international travels.

Global tourism expenditures in 2017 were about USD 3480 billion in domestic tourism, USD 1320 billion in international tourism and USD 830 billion in governmental expenditures on tourism. The total expenditure of tourists was USD 4890 billion; out of which, about 77% was leisure related (USD 3690 billion) and 23% business related[1]; World Bank estimates are somewhat lower. International tourism has grown rapidly. The annual average number of travelers increased by 2.8%,[2] which is more than double from 1995 to 2015; during which time expenditures rise by 3.7% a year implies that they nearly tripled.[3] The direct and indirect global expenditures are even higher because 1 dollar expenditures by tourists generate a global multiplier of 2.6 dollars; it could even generate as much as 3.0 dollars if we include induced products (products and services indirectly linked to tourism, e.g., local retail, transport, souvenirs, and so on). The global USD 290 average expenditure of a tourist has generated USD 760 income, but the countries' multipliers vary because services are often imported.[4]

Ecotourism

Ecotourism covered about 24% of all tourism expenditures in 2015 if based on opinions about it as being the tourist destination aiming at experiences in nature.[5] Meaning of it is that about USD 900 billion is spent on the consumption of bioresources for leisure; other experts estimated these expenditures at USD 600 billion (Balmford et al. 2015). The expenditures are an average of USD 117 per person. Even though the number of visits to nature is disputed because motivations of tourists overlap, and

[1]The world bank data https://knoema.com/atlas/topics/Tourism, accessed 12-4-2018.

[2]https://data.worldbank.org/indicator/ST.INT.DPRT.

[3]https://data.worldbank.org/indicator/ST.INT.XPND.CD.

[4]https://www.wttc.org/-/media/files/reports/economic-impact-research/2017-documents/global-economic-impact-and-issues-2017.pdf (accessed 12-4-2018).

[5]https://sustainabledevelopment.un.org/content/documents/4099Presentation%2001%20Kelly%20Bricker%20-%20full%20presentation.pdf (accessed 12-4-2018).

Table 13.2 Characteristics, visits, and income of select national parks

	I	II	III	IV	V	VI	VII	VIII
Area 1000 km^2	0.04	0.1	3.4	0.08	0.02	1.87	3.2	4.41
Community × 1000		3.5	10.2	10.0	2.0	3.5	90	36.0
Million visits year	0.1	1.0	0.05	0.	0.75	2.2	0.62	21.0
State support in € Mill.	0.2		2.9		0.6	10.0	0.06	
Tourist income in € Mill.	79.3	25.0			0.05	76.3	0.46	
Costs in € Mill.	0.17	2.0	0.69	0.47				

I. Alde Feanen (Netherlands), II. Weerribben-Wieden (Netherlands), III. Hardangervidda (Norway), IV. Lille Vildmose (Denmark), V Söderåsen (Sweden), VI. Loch Lomond (Scotland/United Kingdom), VII. British Waterways (United Kingdom), VIII. Region Uthlande (Germany)
Note: Some data are unavailable as shown by empty boxes

observation methods are imperfect; still, the number of tourists using bioresources increases (Balmford et al. 2009).

Ecotourism constitutes a large and growing value. For instance, it is a few times larger than the global value of biofuels, measured by a high price of USD 100 per barrel of oil equivalents. Ecotourism collides with the use of bioresources that degrades biodiversity, beauty, silence, and other amenities, which implies choices or preferences being made in the planning of land for activities. Despite the large income from ecotourism, as compared to traditional uses of bioresources, decision making is usually disadvantageous because it rivals vested interests in agriculture, forestry, and the like. The challenge within ecotourism is about how to balance pressures on nature caused by the growing number of visitors to nature and the environment. It is possible to strike a balance if distortions by constructions are prevented and tourists receive guidance about how to foster nature (Krozer and Christensen-Redzepovic 2006).

Good natural and environmental qualities generate tourism and income. High expenditures on ecotourism are expressed as high income in many national parks. Table 13.2 shows the number of visitors in a few national parks, their annual income from visits, and government funding in the mid-2000s. The incomes from tourism are substantial when national park authorities take initiative and are allowed to attract visitors within the framework of nature conservation. The income from visitors usually exceeds governmental support, but some national parks are costly to maintain.

Maintaining availability of bioresources for tourism development is an important source of income for many countries. National parks, herewith, can be considered as experimental grounds for the development of a sustainable use of bioresources.

13.4 Outdoor Education

A better understanding of nature is a sine qua non for the sustainable use of bioresources. A prime tool, herewith, is education. This tool is often in the hands of environmental organizations whose membership exceeds 16 million persons in Europe, and 14 million persons in the United States. If we assume that the expenditures of these organizations are largely directed toward education, the global expenditure on education about bioresources would then exceed USD 42 billion a year (Krozer 2015); what seems like a large figure in total boils down to only USD 5.5 average per person.

A Learning Method

Outdoor education engages all senses for participation in nature: audio, visual, taste, touch, and smell. The prime objective is the exposure to ecological qualities which encompass adventure, school camps, wilderness therapy, outdoor tourism along with education about the environment and nature.[6] It addresses the sense of color, shapes, patterns; the behavior of plants and animals in interdependence with soil, water, and air; and, sensitivity of ecosystems to the environment and human activities. In addition, environmental alertness and awareness, resolution of environmental issues, practices for minimum impact, self-reliance in nature, and responsible behaviors can be taught (Ford 1986).

During the last few decades, outdoor education has been enriched with technologies that foster an experience of nature without physical interventions, instead using communication and information technologies. They enable the sensing of bioresources without interference with plants and animals or disturbance of their habitat.[7] This is in addition to the promotion of bioresources, landscape, scenery, environment, cultural heritage, and outdoor facilities (Almer et al. 2002). They also encourage the discovery of environmental qualities, such as darkness and silence. Guided tours foster the engagement of people with nature, for instance, the appreciation of endangered locations. Several technologies used in the national parks are shown in the Appendix.

Tools

Many tools are available, and a few examples illustrate this. 'WhereIGo'[8] is a tool for creating and playing GPS-based adventures in the real world. The use of GPS enables guidance to distant locations. GPS Ranger (tm) and the Zoo Ranger (tm)[9] enable multimedia content using satellite recognition of GPS co-ordinates. They are designed to enhance visitor experience in national parks, to monuments, zoos, and other sites of natural and historic importance and value. General Packet Radio Service (GPRS) and Universal Mobile Telecommunication System (UMTS) enable access to the Internet and the messaging of multimedia information such as pictures, graphics, videos, and audio; as well as voice and data to mobile wireless subscribers. Augmented reality technologies (ART) and virtual reality are interactive visualization systems which combine binoculars with a realistic scene, enhanced with multimedia information

[6]Outdoor Education R&E Centre (2008) What is Outdoor Education? Available at: http://wilderdom.com/definitions/definitions.html (accessed 12-4-2018).

[7].https://sustainabledevelopment.un.org/content/documents/4099Presentation%2001%20Kelly%20Bricker%20-%20full%20presentation.pdf (accessed 12-4-2018).

[8]http://mydigihound.com/?p=178.

[9]http://mydigihound.com/?author=1.

(Fritz et al. 2005). Augmented reality walks simulate scenes of bioresources[10] and Archeoguide simulates walk-through visits of heritage sites. Several national parks in the USA have added Augmented Reality devices that superimpose animations of fossil remains and wildlife in remote areas.

Remote sensing and global information systems are employed to aid explanations of the relationships between species, their habitat, observation of remote areas, and the identification of locations and species. Sensor techniques can visualize nature and stimulate activities outdoors, for instance, to get a sense of qualities or experiences that are not attainable for humans without additional equipment. These can range from seeing forms, hearing voices, sensing smells, visualization, and monitoring of the environment, habitat and wildlife, as well as information about different aspects of people's performance and relaxation in nature.[11]

Connectivity in networks fosters the dissemination of information about the arts, heritage, localization of qualities, and events on predefined trails with limited interventions in the natural processes.[12] For instance, geocache is games that encourage such connectivity about nature. Pressures on nature can be mitigated by the proper use of information and communication technologies in combination with virtual intelligence, gaming, and climate-conditioning for mass tourism at home, meeting points, or in communities.

13.5 Conclusions

Bioresources deliver numerous tangible supportive, regulating, provisioning, and cultural services. The latter are rarely explored, though they generate a high and growing global income. Bioresources in urban areas foster healthy lifestyles and reduce the costs of health care. Experiences of nature generate income from tourism. Education in nature enables the sensing of nature and environmental qualities with the use of modern technologies. Due to the cultural services, bioresources expenditures shift from material-intensive products to material-extensive services and contribute to a better understanding of nature.

Appendix

See Table 13.3.

---

[10]HMD is a display device, worn on the head or as part of a helmet, that has a small display optic in front of one (monocular HMD) or each eye (binocular HMD).

[11]Available: http://www.vti.fi/en/news-events/articles/sporting-activities/.

[12]Sensors can be linked to GSM/sms/mms, GPRS/UMTS, Wifi/Wimax/Bluetooth and other devices used with PDA, smartphone, tablets, videos, cameras. http://www.freepatentsonline.com/20040008660.html.

Table 13.3 Uses of multimedia in national parks

		ICT												
	National parks	GPS guide	PDA	3D tech	Pod casting	Info panels	Web cams	Pictures, movies	SMS	Bluetooth/ wireless Sensors	Video/ audio Tours	Games	Database/ webpage	Virtual reality/ tours
1	Alde Feanen, NL	x				X	x		x			x	x	
2	Weerribben-Wieden, NL	x												
3	Hardangervidda, NO	x	x	x		X							x	
4	Lille Vildmose, DK	x		x		X					x		x	
5	Söderåsen, SE	x			x								x	
6	Loch Lomond & Trossachs, UK		x	x							x		x	
7	Peak District, UK					X		x					x	
8	Kakadu, AU					X		x		x			x	

(continued)

Table 13.3 (continued)

		ICT												
	National parks	GPS guide	PDA	3D tech	Pod casting	Info panels	Web cams	Pictures, movies	SMS	Bluetooth/ wireless Sensors	Video/ audio Tours	Games	Database/ webpage	Virtual reality/ tours
9	Litchfield, AU					X							x	
10	Victoria, AU					X							x	
11	Yellowstone, Wyoming, US[a]	x					x						x	x
12	Grand Teton, Wyoming, US	x					x						x	x
13	Acadia, Maine, US							x			x		x	x[b]
14	Grand Canyon, Arizona, US[c]						x	x			x		x	x
15	Yosemite, California, US						x	x			x		x	

[a]http://www.virtualtourguide.net/

[b]https://imrcms.nps.gov/grca/photosmultimedia/index.htm, http://www.nps.gov/acad/photosmultimedia/index.htm

[c]https://imrcms.nps.gov/grca/photosmultimedia/index.htm

References

Almer A., Zeiner H, Derler C, Stelzl H (2002) National park information system. In: Proceedings of the 16th international conference, informatics for environmental protection. Vienna, Austria, September 2002

Balmford A, Green JMH, Anderson M, Beresford J, Huang C, Naidoo R et al (2015) Walk on the wild side: estimating the global magnitude of visits to protected areas. PLoS Biol 13(2):e1002074. https://doi.org/10.1371/journal.pbio.1002074

Balmford A, Beresford J, Green J, Naidoo R, Walpole M, Manica A (2009) A global perspective on trend in nature-based tourism. PLoS Biol 7(6)

Beute F, de Kort YAW (2013) Salutogenic effects of the environment: review of health protective effects of nature and daylight. Appl Psychol Health Well-Being. https://doi.org/10.1111/aphw.12019

D'Alessandro D, Buffoli M, Capasso L, Fara GM, Rebecchi A, Capolongo S (2015) Green areas and public health: improving wellbeing and physical activity in the urban context. Epidemiol Prev 39(4)Suppl 1: 8–13

Ford P (1986) Outdoor education: definition and philosophy. ERIC Clearinghouse on Rural Education and Small Schools Digest, Las Cruces, NM; Available at: http://wilderdom.com/definitions/definitions.html

Fritz F, Susperregui A, Linaza MT (2005) Enhancing cultural tourism experiences with augmented reality technologies. Asociación VICOMTech, San Sebastian, Spain; The 6th international symposium on virtual reality, archaeology and cultural heritage VAST (2005). Available at: http://public-repository.epoch-net.org/publications/VAST2005/shortpapers/short2005.pdf

Godbey GC, Caldwell LL, Floyd M, Payne LL (2005) Contributions of leisure studies and recreation and park management research to the active living agenda. Am J Prev Med 28:150–158

KPMG (2012) Groen, Gezond en Productief, Ministerie Economisch zaken, Landbouw en Innovatie, Den Haag

Krozer Y (2015) Theories and practices on innovating for sustainable development. Springer, Heildelberg/Dordrecht Chapter 3

Krozer Y, Christensen-Redzepovic E (2006) Sustainable innovations at tourist destinations. Tourism Rev Int 10:113–124

Maller C, Townsend M, Brown P St, Leger L (2002) Healthy parks, healthy people: the health benefits of contact with nature in a park context. Deakin University, Melbourne

Milcu AI, Hanspach J, Abson D, Fischer J (2013) Cultural ecosystem services: a literature review and prospects future research. Ecol Soc 18(3):44. http://www.ecologyandsociety.org/vol18/iss3/art44/

Nieuwenhuijsen MJ, Kruize H, Gidlow C, Andrusaityte S, Antó JM, Basagaña X, Cirach M, Dadvand P, Daneleviciute A, Donaire-Gonzalez D, Garcia J, Jerrett M, Jones M, Julvez J, van Kempen E, van Kamp I, Maas J, Seto E, Smith G, Triguero M, Wendel-Vos W, Wright J, Zufferey J, van de Hazel P-J, Lawrence R, Grazuleviciene R (2014) Positive health effects of the natural outdoor environment in typical populations in different regions in Europe (PHENOTYPE): a study programme protocol. BMJ Open 2014(4):e004951. https://doi.org/10.1136/bmjopen-2014-004951

Triguero-Mas M, Donaire-Gonzalez D, Seto E, Valentín A, Smith G, Martínez D, Carrasco-Turigas G, Masterson D, Berg van de M, Ambròs A, Martínez-Íñiguez T, Dedele A, Hurst G, Ellis N, Grazulevicius T, Voorsmit M, Cirach M, Cirac-Claveras J, Swart W, Clasquin E, Maas J, Wendel-Vos W, Jerrett M, Gražuleviciene R, Kruize H, Gidlow CJ, Nieuwenhuijsen MJ (2017) Living close to natural outdoor environments in four European cities: adults' contact with the environments and physical activity. Int J Environ Res Public Health 14:1162. https://doi.org/10.3390/ijerph14101162

Ulrich RS (1984) View through a window may influence recovery from surgery. Science 224(4647):420–421

Veeneklaas F, Salverda I, Dam van R, During R (2011), Empirisch onderzoek naar de relatie Mens - Natuu/Groen. Een State of Art, Alterra 2198, Wageningen

Vreke J, Donkers JL, Langers F, Salverda IE, Veeneklaas FR (2006) Potenties van groen, Alterra Wageningen

Chapter 14
Conclusions

Yoram Krozer and Michael Narodoslawsky

Economy is comprehended as circular systems of transactions that reflect what people really pay, rather than what they aspire to. Various properties of bioresources are underestimated in the economic systems because they are not expressed in money; for instance the properties that bioresources are non-exhaustible, replenish nature because provide nutrients for new life, store and degradation of pollution and so on. Those ethical attributes that address the care for nature are increasingly demanded even though many highly valued attributes that contribute to welfare have no market prices. Despite this deficiency in all economies, many valuable uses of bioresources refer to the care for nature. Even more opportunities can be found when those benign properties of bioresources are valued in transactions. This leads people to gain knowledge about the various uses of bioresources, and the liabilities for damages to other interests or nature, which is possible if environmental qualities are integrated in all transactions.

Globally, bioresources are abundant for consumption but unfairly divided across income and demographics, as several practices are wasteful; whereas many people lack bioresources too. While the global consumption is about fifth of all materials, it forms only one per cent of the global value. This disparity between the mass and value of bioresources increases. Moreover, the production value of bioresources declines in real terms. Services generate a higher value, in particular knowledge-intensive ones, as opposed to material-intensive ones. Innovating with higher value bioresources and fostering nature for services therefore should be pursued as a goal rather than aiming solely at a higher material output. Regarding many innovative initiatives, there are reasons to be optimistic; particularly when policy support to vested interests shifts

Y. Krozer (✉)
University of Twente - CSTM, Enschede and Sustainable Innovations Academy, Amsterdam, The Netherlands
e-mail: krozer@xs4all.nl

M. Narodoslawsky
Institute of Process and Particle Engineering, Graz University of Technology, Graz, Austria

Y. Krozer and M. Narodoslawsky (eds.), *Economics of Bioresources*,
https://doi.org/10.1007/978-3-030-14618-4_14

towards sustainable innovators in bioresources, and when liabilities for damages are imposed.

Regarding the low value of bioresources and deficient prices of ecosystem services, the conventional strategy is to scale up and intensify the production of bioresources. This strategy, however, evolves to an uneconomic lock-in because prices hardly cover costs. Innovative shifts to more valuable bioresources are risky because they involve costly know-how and implementation, whereas results remain uncertain. Given the intention to avoid a lock-in, innovative strategies are increasingly pursued, such as—the minimization of input costs, introduction of high value specialties, diversification into services and a dedicated supply to niche markets. They emerge from a demand for better qualities and grow—though slowly—due to dedicated supplies.

New businesses in bioresources commence. Many companies aim at the recycling of organic waste for biofuels, which generates income along with better environmental qualities. Some companies invest in innovations that enable the refining of high value specialties along with lower value biofuels. Various business models can be found. The disadvantage of the higher prices of bioresources when compared to their rival alternatives based on minerals and fossil fuels can be met when functional qualities and attributes related to social conventions and one's image are combined into a coherent value proposition. Innovations that link the high functional qualities of products to the ethical attributes of bioresources, mentioned above, increase the chances of success.

Bioresources can be obtained from cultivation (primary sources), where competition for land is encountered. They are also generated from harvest residues (secondary sources), where competition for various uses is faced, specifically between being a nutrient source for fertile land versus uses such as biofuels. They can also be obtained from the discharge of organic materials as waste (tertiary source), if contaminations do not impede their processing. Numerous processing technologies for bioresources can also be found, from conditioning and storage technologies to production, including various conversions and separation technologies for the processing of residues. Regarding the complexity of value chains based on bioresources, the provision of more products and services from the same bioresource is a major way forward, referred to as 'biorefining'.

The technologies and logistics should match the characteristics of bioresources as well as be adapted to the context in which the bioresources are generated. Regarding this dependence on the local context, no single silver bullet is found because technologies and utilization pathways have to be optimally adapted to the local scenario. Currently, the focus of research is the idea of a biorefinery based on various feedstocks that deliver a spectrum of products from high value specialties, down to lower value biofuels. Since the feedstock is bulky, its supply is cost-effective over short distances, therefore favouring a rather small scale of industrial installations and spatially adapted value chains referred to as the 'downscaling of technologies'. Also, some products must be integrated into the regional system of consumption, on account of their low logistic qualities. A special case to consider is the heat generated in many bio-based processes: a product that can be consumed only locally. While planning biorefineries, regional capabilities, distribution pathways and demands should be considered rather than economies of scale.

At the local level, short rotation cultivation based on fast growing plants delivers high yields. Biofuels are produced when harvest is dried with the use of excess heat from other bio-based processes, or inexpensive solar energy and chipped wood at low costs. This production chain enables cost-effective heating in rural areas. Feedstock requires few inputs, and forestry can be developed on marginal land. The storage of CO_2 is an additional advantage. Communities can own the forestry and heating networks, including the essential technologies for wood logging and chipping, which can evolve towards a biorefinery.

While there is little awareness about biofuels among villagers in South East Europe, they have much interest in the development of district heating based on biofuels. Their motivation is to lower the costs of heating, because not many expect jobs and businesses after its implementation. Authorities and citizens can also participate in decision making and implementation when they are equipped with reliable information. District heating based on local bioresources is sufficiently cost-effective for implementation.

There is a high potential for the production of biomethane, from biogenic waste materials and synthetic natural gas, in power-to-gas applications. By 2050, the consumer demand for heating can be met, cost-effectively, with renewable gas. Therefore, biofuels and renewable energy production should expand, logistics for biofuels be improved, plant design tuned to regional capabilities, production technologies be generated, as well as legal and organizational changes implemented. Thus, economic heating based on bioresources requires a long-term perspective on development in the energy sector, as well as in bioresources.

Manure is by far the largest waste stream, en mass, but its processing into biofuel is difficult because manure is wet though contains various minerals. Low cost logistics, as well as drying, separation and digestion technologies enable cost-effective processing; subject to the co-operation of stakeholders in the production chain. Such processing generates biofuels and reduces pollution. The co-operative approach is possible because agricultural business is rooted in the tradition of co-operations. Experiments in regional chains are recommended. The recycling of manure is feasible in many situations, if pollution reduction is awarded by enforcing liabilities for pollution.

Agriculture is presently the main producer of bioresources and the main cause of impacts that pose risks to bioresource production in future. The reason being that it undermines biodiversity and soil qualities which are key natural resources for that production. Innovations are necessary and possible within the framework of conventional agricultural practices. Biodiversity can be maintained by low-input agriculture and system changes towards various forms intercropping, mix of crops and breeds, as well as by 'organic'. Soil fertility can be fostered by closing cycles of minerals and the usage of lighter heavy machinery, which brings down energy use and CO_2 emissions.

In addition to its present use as a biofuel plants', lignin can be used for various high value products. Its present high value uses are as expanders in lead-acid batteries, flocculants, emulsion stabilizers and water reducing agents. Various uses of lignin that are currently being researched become replacement for phenol-formaldehyde (PF) resins, can be used as carbon fibres in aircrafts and antimicrobe agents, in chemical vanillin and thermal sensors. Modified lignin can reinforce materials, be used as a coating in packaging, as UV protectors and degradable thermoplastic. Many options have been and are being investigated; however, prices are not competitive with the synthesizing of mineral-based compounds. They can be competitive when ethical attributes are valued.

Bioresources deliver tangible ecosystem services of all types—supportive, regulating, provisioning and cultural. The cultural services generate a high and growing global income. Bioresources in urban areas contribute to the prevention of health risks and shorten the hospitalization period. Experiences of nature generate a large income from tourism, without the destruction of nature. Education can foster an understanding wherein nature and environmental qualities are experienced with the use of modern technologies. Those cultural services generate income based on bioresources and foster nature as well.

The development of bioresources for various products, including biofuels, is economically attractive when—the production chain is aligned with regional conditions and demands; ethical attributes of bioresources are valued; and, liabilities for waste disposal and pollution are strictly imposed on market transactions.

Printed by Printforce, the Netherlands